4-21

Probability and Mathematical Statistics (Continued)

SEBER • Linear Regression Analysis
SEBER • Multivariate Observations
SEN • Sequential Nonparametrics: Invariance
Inference
SERFLING • Approximation Theorems of Mathematical Statistics
TJUR • Probability Based on Radon Measures
WILLIAMS • Diffusions, Markov Processes, and Martingales, Volume I: Foundations
ZACKS • Theory of Statistical Inference

Applied Probability and Statistics

ABRAHAM and LEDOLTER • Statistical Methods for Forecasting
AGRESTI • Analysis of Ordinal Categorical Data
AICKIN • Linear Statistical Analysis of Discrete Data
ANDERSON, AUQUIER, HAUCK, OAKES, VANDAELE, and WEISBERG • Statistical Methods for Comparative Studies
ARTHANARI and DODGE • Mathematical Programming in Statistics
BAILEY • The Elements of Stochastic Processes with Applications to the Natural Sciences
BAILEY • Mathematics, Statistics and Systems for Health
BARNETT • Interpreting Multivariate Data
BARNETT and LEWIS • Outliers in Statistical Data
BARTHOLOMEW • Stochastic Models for Social Processes, *Third Edition*
BARTHOLOMEW and FORBES • Statistical Techniques for Manpower Planning
BECK and ARNOLD • Parameter Estimation in Engineering and Science
BELSLEY, KUH, and WELSCH • Regression Diagnostics: Identifying Influential Data and Sources of Collinearity
BHAT • Elements of Applied Stochastic Processes
BLOOMFIELD • Fourier Analysis of Time Series: An Introduction
BOX • R. A. Fisher, The Life of a Scientist
BOX and DRAPER • Evolutionary Operation: A Statistical Method for Process Improvement
BOX, HUNTER, and HUNTER • Statistics for Experimenters: An Introduction to Design, Data Analysis, and Model Building
BROWN and HOLLANDER • Statistics: A Biomedical Introduction
BROWNLEE • Statistical Theory and Methodology in Science and Engineering, *Second Edition*
CHAMBERS • Computational Methods for Data Analysis
CHATTERJEE and PRICE • Regression Analysis by Example
CHOW • Analysis and Control of Dynamic Economic Systems
CHOW • Econometric Analysis by Control Methods
COCHRAN • Sampling Techniques, *Third Edition*
COCHRAN and COX • Experimental Designs, *Second Edition*
CONOVER • Practical Nonparametric Statistics, *Second Edition*
CONOVER and IMAN • Introduction to Modern Business Statistics
CORNELL • Experiments with Mixtures: Designs, Models and The Analysis of Mixture Data
COX • Planning of Experiments
DANIEL • Biostatistics: A Foundation for Analysis in the Health Sciences, *Third Edition*
DANIEL • Applications of Statistics to Industrial Experimentation
DANIEL and WOOD • Fitting Equations to Data: Computer Analysis of Multifactor Data, *Second Edition*
DAVID • Order Statistics, *Second Edition*
DAVISON • Multidimensional Scaling
DEMING • Sample Design in Business Research
DILLON and GOLDSTEIN • Multivariate Analysis: Methods and Applications
DODGE and ROMIG • Sampling Inspection Tables, *Second Edition*
DOWDY and WEARDEN • Statistics for Research

continued on back

INTRODUCTION TO COMBINATORIAL THEORY

INTRODUCTION TO COMBINATORIAL THEORY

R. C. BOSE and B. MANVEL
Colorado State University

JOHN WILEY & SONS
New York · Chichester · Brisbane · Toronto · Singapore

Library of Congress Cataloging in Publication Data:

Bose, R. C. (Raj Chandra), 1901–
Introduction to combinatorial theory.

(Wiley series in probability and mathematical
statistics. Probability and mathematical statistics,
ISSN 0271–6232)
Includes index.
1. Combinatorial analysis. I. Manvel, B. (Bennet)

1943– . II. Title. III. Series.
QA164.B67 1984 511′.6 83–17019
ISBN 0–471–89614–4

Printed in the United States of America

10 9 8 7 6 5 4 3 2 1

Preface

Combinatorial theory is rapidly assuming a position of major importance in mathematics. Combinatorial methods are particularly relevant in statistics and computer science, and the pure mathematics involved is intuitive and appealing.

This book introduces a great variety of combinatorial topics at a level suitable for undergraduates. Other texts emphasize graphs or counting or designs. We introduce all of these areas, as well as finite geometries, SDR's, optimization, and others, displaying an unusual breadth of coverage. Obviously, such breadth requires careful selection of topics. We have selected from the material in each area topics that are basic, useful, typical, and appealing. The treatment is kept as simple as possible, with notation standard and minimal.

The subjects covered fall naturally into four unequal classes. Chapters 1, 2, and 3 survey elementary counting techniques, with applications to probability. Chapter 4 is a brief introduction to graph theory, which only suggests the richness of that very active area. Chapters 5, 6, 7, and 8 introduce constructive combinatorics in an elementary but fairly complete way. Finally, Chapters 9 and 10 touch on problems of choice and optimization, so important to modern applied combinatorics. Preliminary versions of this book have been used for several years at Colorado State University in a course serving undergraduate students in mathematics, computer science, statistics, and engineering. In our experience, it is possible to cover about two-thirds of this material in a one-semester, three-hour-per-week course.

We are grateful to the many people (mostly unnamed) who discovered the lovely mathematics we present here. Thanks are also due to colleagues (particularly Richard Games and Robert Liebler) and students who found errors and offered valuable advice, and Kristy Lahnert, who transformed scribblings into a typescript with remarkable skill and patience. Finally, we would like to acknowledge our debt to Bea Shube, our editor at Wiley, without whose persistent encouragement this book would not have been possible.

R. C. Bose
B. Manvel

Fort Collins, Colorado
November 1983

Contents

INTRODUCTION TO COMBINATORIAL THEORY

CHAPTER 1

Permutations and Combinations

How many different states are possible for a computer byte containing six binary bits? In how many ways can five couples be seated around a table, with sexes alternating and no one sitting next to his or her partner? How many divisors does the number 858 have? In a state with 1,285,000 motor vehicles, what is the most efficient choice of letters and digits for use on the state's license plates?

Questions such as these are the subject of this chapter. We will be counting choices or distributions which may be ordered or unordered and in which repetition may or may not be allowed. The tools we develop are simple and very general, so they can be used to solve a great variety of problems. The difficult part of solving such problems of counting is almost always figuring out just which principles are needed and how they should be applied. Thus, the majority of your efforts in studying this chapter should be directed at a careful reading of the examples and diligent efforts at solving as many problems as possible.

1.1. TWO COUNTING PRINCIPLES

The basic idea in the theory of counting is to reduce a large problem to smaller ones. Instead of counting elements in a set, count elements in subsets. Instead of listing all actions, list representative actions. Instead of finding the number of ways of doing a task, find the number of ways of doing various sub-tasks. Such reductions are useful because of two basic principles.

The *addition principle* says that if one task can be performed in m ways and another task in n ways, then one task or the other can be performed in $m + n$ ways. The *multiplication principle* says that if one task can be performed in m ways and then another task can be performed in n ways, the pair of tasks, first one and then the other, can be performed in $m \cdot n$ ways. These obvious and harmless sounding principles are the foundation of enumerative combinatorics, and are applicable in a huge variety of problems.

Example 1. If a license plate consists of a letter and a digit or a digit and a letter, how many different license plates are possible? Here we know there are 10 digits (0, 1, 2, 3, 4, 5, 6, 7, 8, 9) and 26 letters. We attack the problem by breaking the job of constructing a license plate into small tasks. To make a license plate we must first decide if we want to begin with a letter or with a digit. Considering that choice as giving us two separate jobs, we can count the ways to do each one and add the resulting numbers, using the addition principle to find the total number of plates. But if we begin with a letter, we can choose the letter in 26 ways and then choose a digit in 10 ways. Thus, there are $26 \cdot 10 = 260$ ways to make a plate beginning with a letter, by the multiplication principle. Since there are clearly also 260 plates that begin with a digit, we have $260 + 260 = 520$ plates in all.

Both the addition principle and the multiplication principle can be extended to more general statements. The obvious generalization for addition states that if task i can be performed in m_i ways then one of k tasks can be performed in $m_1 + m_2 + \cdots + m_k$ ways. The generalized multiplication principle says that if one task can be done in m_1 ways, a second in m_2 ways, and so until the kth task can be done in m_k ways, then the succession of tasks, one after the other, can be done in exactly $m_1 \cdot m_2 \cdot m_3 \cdots m_k$ ways.

Note that the multiplication principle applies to situations in which one performs several tasks, successively. In situations involving only a single choice or task from a range of possibilities the addition principle is used.

Example 2. A club with 12 members is to elect a president, a vice-president, a secretary, and a treasurer. How many different outcomes are possible? There are 12 ways the first task (selecting a president) can be done, and then 11 ways to select a vice-president, 10 to select a secretary, and 9 a treasurer, and finally there are $12 \cdot 11 \cdot 10 \cdot 9 = 11{,}880$ ways to do the whole job of selection, corresponding to 11,880 possible outcomes.

Example 3. If a red die and a green die are thrown, how many different outcomes are possible? Each die must show one of the numbers 1 through 6. Thinking of the throwing of the two dice as separate tasks, we see the first task (throw the red die) can end in one of 6 outcomes, and the second task (throw the green die) in a similar number. Thus there are $6 \cdot 6 = 36$ possible outcomes if both are thrown.

Example 4. The won–lost record determines the ranking of the teams in a football league. If we assume that no two teams have equal records, how many different rankings are possible in a league with eight teams? Here we can choose the team which is first in 8 ways, the second team in 7 ways, and so on, until there is only one team left to be in last place. By the multiplication principle, the entire ranking job can be performed in $8 \cdot 7 \cdot 6 \cdot 5 \cdot 4 \cdot 3 \cdot 2 \cdot 1 = 40{,}320$ ways.

EXERCISES

Note. In these exercises, and throughout the text, the wording is literal rather than colloquial. Thus "two letters" means two not necessarily distinct letters. If distinct objects must be used, that will be stated.

1. If a classroom contains five rows, each with seven chairs, how many chairs are in the room? What counting principle is being used here?
2. If a license plate contains two digits followed by two letters, how many different plates are possible?
3. If the license plates in exercise 2 can contain no repeated digits or letters, how many plates are possible?
4. License plates consist of one or two letters followed by one, two, or three digits. How many different plates are possible?
5. A mail-order company sells five styles of men's slacks. Each style is available in ten lengths, eight waist sizes, and four colors. How many different kinds of slacks does the company have to stock?
6. A clothing company keeps in stock 1236 kinds of shirts, 13,280 kinds of shoes, and 540 kinds of socks. How many kinds of clothing does the company stock?
7. How many different positive integers containing no digits besides 7, 8, and 9 contain each of those digits exactly once? At most once?
8. John has a penny, a nickel, a dime, and a quarter. He must distribute them to six friends. In how many ways can he do that if (a) no friend gets more than one coin, (b) some friends may get more than one coin, (c) some friend must get more than one coin?

9. If n couples are at a dance, in how many ways can the men and women be paired up for a single dance?

10. If n has prime decomposition $n = p_1^{a_1} \cdot p_2^{a_2} \cdot \ldots \cdot p_k^{a_k}$, then show that n has exactly $(a_1 + 1) \cdot (a_2 + 1) \cdot \ldots \cdot (a_k + 1)$ factors, including 1 and n. Use this to show that every square has an odd number of factors, and every nonsquare has an even number of factors.

1.2. ORDERED CHOICES

We have seen that various problems can be viewed as involving the choice of r objects from a set of n objects. The number of ways of making such a choice depends on whether or not a given object may be selected more than once and whether or not the order of choice is important. In this section we will consider ordered choices, with and without repetition being allowed. The more difficult problem of unordered choice will be covered in Section 1.3.

An ordered choice of r not necessarily distinct elements from a set S is called an *r-sample* of S.

Theorem 1.1. The number of r-samples from a set with n elements is n^r.

Proof. We count the number of ways of forming an r-sample using the multiplication principle. The first element may be chosen in n ways, the second in n ways, and so on, r times, so there are $n \cdot n \cdot n \cdot \cdots \cdot n = n^r$ ways to do the whole task of choosing an r-sample.

Example 5. A football team wears jerseys that each display a pair of digits. How many different jerseys are possible? We clearly want a 2-sample from the set of digits for each jersey, so there are $10^2 = 100$ possibilities.

Example 6. How many five letter "words" can be formed using the English alphabet? The number of 5-samples of a 26-set is $26^5 = 11{,}881{,}376$.

Example 7. How many states are possible for a computer byte containing six binary bits? Since each bit can take on two different values, we are dealing with 6-samples of a 2-set. Thus, the number is $2^6 = 64$.

Example 8. How many 2-scoop ice cream cones are possible, using 31 flavors? Some care must be taken here, because it is not immediately obvious (as it was in the preceding examples) that this is an ordered rather than an unordered sampling of two things from 31. Let us suppose though, that we adopt the gourmet viewpoint that a chocolate and vanilla cone is in no sense the same as a vanilla and chocolate one. In that case we are dealing with 2-samples of a 31-set, and the number of possible cones in $31^2 = 961$.

Suppose now that in the preceding example we restrict ourselves by insisting that the two flavors chosen for our ice cream cone be different. Clearly then we are merely disallowing the 31 cones that are of a single flavor each, and the number of possible cones becomes $31^2 - 31 = 930$. We now examine what happens in general if repetition is forbidden.

Any r-sample from a set S in which the elements chosen are distinct is called an *r-permutation* of S. If S has exactly r elements, an r-permutation of S is called simply a *permutation* of S.

Theorem 1.2. The number of r-permutations of an n-set is

$$n_r = n(n-1)(n-2)\cdots(n-r+1). \tag{1.1}$$

Proof. Again we use the multiplication principle, noting now that the first choice is from n elements, the second from $n-1$, and so on. The rth choice is from only $n-(r-1)$ elements, so we obtain (1.1).

If r is small, formula (1.1) is convenient. For large r, however, it is convenient to express the numbers n_r in factorial notation. If n is a positive integer the number $n!$, read *n factorial*, is the product of all integers from n down to 1. Thus $5! = 5 \cdot 4 \cdot 3 \cdot 2 \cdot 1 = 120$ and $3! = 3 \cdot 2 \cdot 1 = 6$. With factorial notation in mind, we can rewrite equation (1.1) as

$$n_r = \frac{n(n-1)(n-2)\cdots 2 \cdot 1}{(n-r)\cdots 2 \cdot 1} \qquad (\text{for } r < n).$$

Thus

$$n_r = \frac{n!}{(n-r)!}. \tag{1.2}$$

If $n = r$, equation (1.1) says we should have $n_r = n!$. In view of this fact and formula (1.2), we find it convenient to define $0!$ to be equal to 1. Notice also that equation (1.2) makes sense for $r = 0$ (if we set $n_0 = 1$) while equation (1.1) does not. Thus, for convenience, we let $n_0 = 1$ for all positive integers n.

Since n_r is the number of ordered choices of r distinct elements from an n-set, formulas (1.1) and (1.2) enable us to solve many problems with minimal effort.

Example 9. Recall that in Example 2 we were to select four officers for a club with twelve members. This is clearly an ordered choice of four elements from a 12-set, so the answer is $12_4 = 12!/(12-4)! = 12 \cdot 11 \cdot 10 \cdot 9 = 11{,}880$. Notice that we do not calculate 12! or 8!, but use cancellation to simplify the arithmetic involved.

Example 10. In how many ways can the letters A, B, C, and D be arranged in a row? Here we are making ordered choices of four objects from four, so we need only count the number of permutations of the letters A, B, C, and D. There are $4_4 = 4!/(4-4)! = 4! = 24$ such permutations.

Example 11. A luggage company applies 3-letter monograms to suitcases by transfer from plastic decals. How many different 3-letter decals must they stock to be able to handle any monogram? At first glance, one might think the answer is 26_3 but this is not correct since repetition of letters is allowed. We are, in fact, counting samples rather than permutations, so the correct answer is $26^3 = 17{,}576$.

EXERCISES

1. Evaluate 5_4, 6_3, and 50_2.
2. Bill has nine members on his baseball team. In how many ways can he choose players for the four infield positions (first base, second base, third base, and shortstop)?
3. How many numbers between 100 and 999 (inclusive) have distinct digits?
4. How many "words" can be formed using the letters of STOP (each at most one time): (a) if all four letters must be used; (b) if some (or all) of the letters may be omitted?
5. How many 4-digit numbers contain no zero? How many of those have distinct digits?
6. How many arrangements are there of a deck of 52 cards?
7. How many "words" can be made using four A's and one B?
8. Bob is constructing 6-letter codes from metal letters. (a) If he has 3 A's, 2 F's, and an H, how many codes can he make? (b) He finds 4 more A's, 7 more F's, and 6 more H's in his garage. Now how many different codes can he make?
9. How many integers greater than 6600 have distinct digits, not including the digits 7, 8, or 9?

10. Show that $n_n = n_{n-1}$ for $n \geqslant 1$.

11. Show that $n_r (n - r)_k = n_{r+k}$.

12. An efficient algorithm for generating all permutations of the numbers 1 through n is illustrated below for $n = 2$ and 3.

$n = 2$:

1			1	2
1	→	2	1	

$n = 3$:

1 2			1		2	3
1 2			1	3	2	
1 2		3	1		2	
2 1	→	3	2		1	
2 1			2	3	1	
2 1			2		1	3

This can be explained as:

(a) Write down all permutations of 1 to $(n - 1)$, listing each permutation n times.

(b) Place an n in each permutation, in a zig-zag pattern.

For this algorithm:

(i) Generate the permutations of 1 to n, for $n = 4$ and $n = 5$, by hand.

(ii) Write a computer program to print the permutations of 1 to n for $n = 9$.

1.3. UNORDERED CHOICES

The counting of unordered choices is more difficult than the same problem for ordered choices because the multiplication principle essentially dictates an order of choice and thus cannot be applied directly. Using a small trick, involving a double application of the multiplication principle, we will be able to count unordered choices.

An unordered choice of r distinct elements from a set S is called an *r-combination* of S. (Note that an r-combination is just an r-subset of S.) We denote the number of r-combinations of an n-set by $\binom{n}{r}$.

Theorem 1.3. The number of r-combinations of a n-set is

$$\binom{n}{r} = \frac{n!}{r!(n-r)!}. \tag{1.3}$$

Proof. We can divide the collection of n_r r-permutations of an n-set into classes, putting two permutations in the same class if they contain the same r elements. By the definition of $\binom{n}{r}$ there are exactly $\binom{n}{r}$ classes in such a division. Moreover, each of these classes will contain exactly $r!$ permutations since each contains all possible ordered arrangements of a given set of r elements. Thus by the multiplication principle $n_r = \binom{n}{r}r!$. Solving this equation for $\binom{n}{r}$ and inserting the expression in formula (1.2) for n_r we have equation (1.3).

The numbers $\binom{n}{r}$ are known as *binomial coefficients*, and we will study them in detail in Chapter 2. For now, we find them very useful in many counting problems.

Example 12. How many committees of four people can be chosen from a club with twelve members? This question differs from the choice of four officers, in Examples 2 and 9, because the committee selected does not depend on order of selection. Thus, we are making an unordered choice of four elements from a 12-set, and the answer is $\binom{12}{4} = 12!/4!(12-4)!$ $= 12 \cdot 11 \cdot 10 \cdot 9/4 \cdot 3 \cdot 2 \cdot 1 = 495$. Note again the use of cancellation to simplify the computation.

Example 13. If Bill wants to take two math courses and two history courses, and there are five suitable math courses and four suitable history courses available, in how many ways can he choose the four courses? He can choose the math courses in $\binom{5}{2} = 10$ ways and the history courses in $\binom{4}{2} = 6$ ways, so he can do both things together in $10 \cdot 6 = 60$ ways.

Example 14. In how many ways can seven dashes and five slashes be arranged in a row? Since we are counting ordered arrangements, it is somewhat surprising that the answer is $\binom{12}{5}$. To see that, view each arrangement as one possible way of filling twelve boxes, seven with dashes and five with slashes, for example as

To do this, we first choose the five boxes in which slashes are to appear, which can be done in $\binom{12}{5}$ ways, and then fill the remaining boxes with dashes. The number obtained if the seven dashes are placed first is $\binom{12}{7}$. Of course, these two numbers are equal since from (1.3)

$$\binom{12}{5} = \binom{12}{7} = \frac{12!}{5!\,7!}.$$

Example 15. How many words can be formed using the letters of MISSISSIPPI?

There are four S's, two P's, four I's, and a single M for a total of eleven letters. If there are eleven boxes, then we can first place the four S's in $\binom{11}{4}$ ways. Then the two P's can be placed in the seven unoccupied boxes in $\binom{7}{2}$ ways. The four I's can now be placed in the five unoccupied boxes in $\binom{5}{4}$ ways, and finally there is only one way of placing M in the single box remaining. Thus the required number of ways is

$$\binom{11}{4}\binom{7}{2}\binom{5}{4}\binom{1}{1} = \frac{11!}{4!\,7!}\cdot\frac{7!}{2!\,5!}\cdot\frac{5!}{1!\,4!}\cdot\frac{1!}{1!\,1!}$$
$$= \frac{11!}{4!\,2!\,4!\,1!}.$$

These two examples suggest the following theorem which can be proved by counting the number of ways as in the examples.

Theorem 1.4. Given n_i objects of type i, for $i = 1, \ldots, k$, the total of $n = n_1 + n_2 + \cdots + n_k$ objects can be arranged in a row in

$$\binom{n}{n_1, n_2, \ldots, n_k} = \frac{n!}{n_1!\,n_2!\cdots n_k!} \tag{1.4}$$

ways.

The numbers $\binom{n}{n_1, n_2, \ldots, n_k}$ in the theorem are called *multinomial coefficients*, and we shall discuss them more thoroughly in Chapter 2. Note that for the case $k = 2$

$$\binom{n}{n_1, n_2} = \binom{n}{n_1} = \binom{n}{n_2}$$

since $n = n_1 + n_2$.

We are now ready to count unordered choices with repetitions allowed. A choice of r not necessarily distinct elements from a set S is called an *r-selection* of S. Clearly an r-selection in which the elements chosen are distinct is an r-combination. In order to count r-selections we place such selections in a one-to-one correspondence with a set we have already counted.

Theorem 1.5. The number of r-selections of an n-set is $\binom{r+n-1}{n-1}$.

Proof. Suppose $S = \{s_1, s_2, \ldots, s_n\}$. *Each* r-selection of S corresponds to a sequence of r dashes and $n - 1$ slashes, where the dashes indicate presence of an element and the slashes indicate a change from one element to the next (in the subscript ordering we have specified). The following examples illustrate this in a case where $n = 4$ and $r = 7$.

r-Selection	Dash–Slash Sequence
$s_1s_1s_2s_3s_3s_3s_4$	– – / – / – – – / –
$s_1s_1s_2s_2s_2s_4s_4$	– – / – – – / / – –
$s_2s_2s_2s_2s_2s_2s_3$	/ – – – – – – / – /
$s_1s_1s_2s_2s_2s_2s_2$	– – / – – – – – / /

Note that if some elements are absent from the beginning then we put one slash for each of them in the beginning. Similarly, if some elements are absent from the end we put one slash for each of them at the end.

Since this correspondence is clearly one-to-one, the number of r-selections of an n-set is the same as the number of dash–slash sequences with r dashes and $n - 1$ slashes. By Theorem 1.4, that number is $\binom{r+n-1}{n-1}$.

Example 16. The luggage maker in Example 11 decides that he cannot stock 17,576 different 3-letter monograms, so he changes his procedure. Instead of having monograms made up as triples of letters, he has them prepared as small packets, with three separate letters not necessarily distinct in each packet. The person applying the monogram selects a packet with the correct three letters, arranges them in order, and applies them. How many types of packets must be stocked? Here we have a 26-set (the alphabet) and we are counting 3-selections, since order is unimportant and repetition is allowed. By Theorem 1.5 the answer is $\binom{3+26-1}{25} = \binom{28}{25}$ $= 28!/25!3! = 28 \cdot 27 \cdot 26/3 \cdot 2 \cdot 1 = 3276$. This is a considerable improvement, but perhaps he should just stock 26 kinds of letters! Note also that the answer is not the number of ordered monograms ($26^3 = 17{,}576$) divided by 3!, since triples with repeated letters do not have 3! different orderings.

Example 17. Tommy can choose from five kinds of penny candies, and he has a dime to spend. How many different selections of candy are possible for him? Here we have unordered choice, with repetition allowed, so the answer is $\binom{10+5-1}{5-1} = \binom{14}{4} = 1001$. No wonder it takes him so long to choose!

EXERCISES

1. Find $\binom{20}{3}$, $\binom{10}{2}$, and $\binom{10}{8}$.
2. Show $\binom{n}{r} = \binom{n}{n-r}$ for all n and r.
3. Find $\binom{15}{5,3,2,5}$ and $\binom{15}{5,5,3,2}$.

4. Mary needs to take five courses next semester to keep her scholarship. If there are nine courses she may take, in how many ways can she choose five courses?

5. Sport-Pro is organizing an International Asparagus-Cutting Competition for eight teams. If eleven teams want to compete, in how many ways can the competing teams be chosen?

6. If a bakery has five kinds of cookies, in how many ways can a dozen be chosen?

7. How many "words" can be made using all the letters of "MATHEMATICS"?

8. Jack has six toys and wants to trade two toys with Jim, who has eight toys. In how many ways can they trade?

9. In how many ways can one card of each suit be chosen from a standard 52-card deck?

10. Six fullbacks and eight linebackers are available to be drafted. Dallas is to choose one player and then Denver wants to choose a fullback and a linebacker. What type of player should Dallas choose to leave Denver with the fewest choices?

11. (a) If one-fourth of the 3-subsets of $\{1, 2, \ldots, n\}$ contain n, what is n?
    (b) What fraction of the r-subsets of an n-set contain a given element?

12. The central library in Slobovia contains dictionaries for the direct translation of any one of the country's seventeen dialects into any other. For example, there is a dictionary for translating back and forth between Wimballa and Shimtuli. How many dictionaries are there?

1.4. CIRCULAR AND RESTRICTED ARRANGEMENTS

In some sense our development in the last two sections has been unnecessary, since all the problems in this chapter can be solved by sufficiently clever application of the two fundamental counting principles. We suspect, however, that most people would have a hard time counting, say, unordered choices if they had not been shown the trick. We cannot hope to give examples of every type of counting problem that may arise, together with the methods for their solutions, but there are several other standard counting tricks that are required frequently enough to deserve comment. Instead of listing theorems, we merely present examples that display typical tricks.

Example 18. In how many ways can the numbers 1 to 100 be arranged around a roulette wheel? We can solve this problem directly, using the multiplication principle. Suppose we begin with a blank wheel and paint on the numbers 1 to 100, beginning with 1, around the rim of the wheel. Each number should go in one of the 100 equal-sized spaces around the rim, to make the game fair. It may seem that there are 100 different places that the 1 may be placed, but that is not so. Since the wheel is free to rotate, all 100 positions are equivalent, so there is only one way to place the 1. Once the 1 has been placed, however, every subsequent number will take its position relative to the 1, so there are 99 places to use for the 2, 98 for the 3, and so on. Thus, the answer to the problem is $99! \cong 2.76 \cdot 10^{129}$

Example 19. In how many ways can four couples be seated at a circular table, alternating by sex? Here again, we assume that the table can rotate, so that if one arrangement can be rotated to another, the two are the same. The solution involves good manners—we seat the ladies first! There is essentially one way to seat the first lady, then the other ladies can be placed (in alternating seats) in 3! ways. This leaves four spaces for the men, who can be seated in 4! ways. Thus, the answer is $3!4! = 144$.

Example 20. How many 5-letter words can be formed that contain a vowel? This rather difficult problem is reduced to an easy one if we note that the number of 5-letter words is simply 26^5 and the number of such words that contain *no* vowel is simply 21^5. Thus the answer to the stated problem is just $26^5 - 21^5$.

In general, if it is hard to count objects with a given property, it may be easier to count the objects that *do not* have the property, and then subtract from the total number of objects to find the number required.

Example 21. How many 2-card blackjack deals (one card up, one card down) contain exactly one ace? Here we can count the total number of 2-card deals ($52 \cdot 51$) and the number of such deals without an ace ($48 \cdot 47$). Thus there are $52 \cdot 51 - 48 \cdot 47 = 396$ deals that contain at least one ace. Since there are $4 \cdot 3 = 12$ deals that contain two aces, there are 384 with exactly one ace. Alternatively, we count those deals with an ace down and a non-ace up as $4 \cdot 48 = 192$, find 192 also for the number of hands that are non-ace, ace, and arrive at the 384 more directly.

Example 22. How many ways can six girls and three boys be lined up so that no two boys are together? This problem is best solved by placing the girls first, which can be done in 6! ways. Each boy must then be placed between two of the girls or at one end of the line, with no two boys in the same spot. There are seven places for boys, so the boys can be placed in $7_3 = 7 \cdot 6 \cdot 5 = 210$ ways. Thus the total number of line-ups is $6! \cdot 210 = 151{,}200$.

Example 23. In how many ways can three numbers be chosen from the numbers 1 to 9 so that no two consecutive numbers are chosen? This problem is quite similar to the previous one, if it is viewed in the following way. We can represent a given choice of numbers by a sequence of 0's and 1's, with 0 indicating that a number is not chosen and 1 indicating that it is. Thus, for example, 100101000 represents the choice of the numbers 1, 4, and 6. The condition that consecutive numbers not be chosen translates to a ban on consecutive 1's. Placing the 0's first, we see that this can be done in exactly one way (unlike girls, 0's are indistinguishable). The three 1's can then be placed into the five places between 0's or at the end or beginning, making seven possible locations in all. There are $\binom{7}{3}$ ways to select three places to receive 1's, so there are $1 \cdot \binom{7}{3}$ different sequences in which no two 1's are consecutive, and that is also the answer to our original problem. Thus the required number is 35.

EXERCISES

1. In how many ways can five A's and seven B's be lined up so that no two A's are adjacent?
2. In how many ways can the letters, A, B, C, D, E, F, G, H, I, J, O, and U be lined up so that no two vowels are together? So that no two consonants are together?
3. How many k-digit numbers are there in the usual (base 10) number system? In the base n number system? (Note that the first digit of a k-digit number must be nonzero.)
4. The Morse code is made up of marks called dots and dashes. "Q", for example, is – – · –. Is it possible to make up such a code so that every letter of the alphabet is represented by at most three marks? At most four?
5. If six identical apples and four identical oranges are to be placed in two identical bowls, five fruits to a bowl, in how many ways can that be done? What if each bowl must contain at least one of each kind of fruit?

6. In how many orders can six people be seated at a round table if one of the people hates one of the other five people and refuses to sit beside him or her?

7. Three of the time slots 8, 9, 10, 11, 12, 1, 2, 3, and 4 are to be selected for calculus classes. In how many ways can that be done if no two classes are to be scheduled in consecutive time slots?

8. If a committee of ten people contains four women, in how many ways can a subcommittee of five be chosen if that subcommittee must by law contain at least one woman?

9. From ten people seated around a table, four are to be chosen to represent the group. Since friends generally sit together it is decided to make the selection in such a way that no two people adjacent to each other are chosen. In how many ways can such a selection of four people be made?

10. In how many permutations of the 26 letters of the alphabet do the letters a and b *not* appear together? In how many do a, b, and c not appear together? Can you generalize this?

1.5. FINITE PROBABILITY

The counting methods we have developed are frequently applied in the theory of probability. That theory has important applications in economics, insurance, medicine, physics, politics, and many other fields, but we shall only develop enough of the subject to show how counting techniques are used.

Probability theory is important because we generally have incomplete information about what will happen in the future. Using partial information we can make an educated guess about what will happen, and it is frequently useful to be able to express numerically the level of confidence that we have in such a guess. Thus, several years ago the U.S. weather service changed from saying, "It is likely to rain" to saying "There is a 60% probability of precipitation". The use of numbers brings precision, where language is inexact. All of our actions are based on mental calculations of the likely and unlikely consequences of each possible action. Of course, we do not usually assign numerical measures to each likelihood, but our minds are automatically weighing various alternatives. If forced to assign values to each possibility, how would we do it, and what would the numbers mean?

When numbers are used to measure likelihood, it seems desirable that an event that is more likely should get a larger number. Furthermore, it seems clear that the numbers we use as probabilities should be bounded above and below. This is so since there are some events that are certain to occur and others that cannot occur at all. It seems reasonable to assign those events that are certain (and therefore most likely of all) the highest number in our system while those that cannot occur get the lowest number. The numbers 1 and 0 are used for these bounds. Thus 1 and 0 are the probabilities of a certain event and an impossible event, respectively, and all other probabilities fall between 1 and 0. What does a specific probability between 0 and 1 express?

Suppose we flip a coin. It is easy to say that the probability it will come up heads is 1/2, but a bit harder to say what that means. Of course, it does not mean that on this flip it will come up one way or the other; that we cannot predict. In fact the information conveyed by the probability is of a long-term nature: If we flip the coin many times, it will come up heads about half of the time. We cannot say it will alternate heads or tails, or that it will come up heads exactly half the time, but it is likely that the outcomes will be split about half and half in the long run. If the coin came up heads 75 times out of 100 we would be surprised and might examine the coin to see if something was special about it. Where does the probability 1/2 come from? Clearly there are two things that can happen when we flip a coin: heads or tails, which seem to be equally likely. Thus we say each will happen half of the time, giving us our probability. In order to understand why this is justified, and how similar methods can be applied to more complex problems, we need some definitions.

A *sample space* of an experiment is the set of all outcomes. Here an outcome is, for example, a head or a tail. In more complicated experiments the outcomes can sometimes be chosen in various ways.

For example, if we throw two dice, we can describe the outcome by saying exactly what number came up on each die (e.g., "die 1 has a 6, die 2 has a 3") or by specifying the total (e.g., "we rolled 9"). The first of these two ways turns out to be preferable, since it gives outcomes that are equally likely, while the other does not. More on this later.

An *event* is a set of possible outcomes. Thus events are just subsets of the sample space. If we flip a coin twice the sample space S is HH, HT, TH, TT. There are sixteen possible events $\varnothing$, {HH}, {HT}, {TH}, {TT}, {HH, HT}, {HH, TH}, {HH, TT}, {HT, TH}, {HT, TT}, {TH, TT}, {HH, HT, TH}, {HH, HT, TT}, {HH, TH, TT}, {HT, TH, TT}, {HH, HT, TH, TT}.

Tom and Harry play the following game. Tom flips a coin twice and wins a dollar if there is at least one head, and loses three dollars if there is no head. Then the event that describes Tom's winning a dollar is {HH, HT, TH}, and the event that describes Tom's losing three dollars is {TT}.

We denote the number of elements in a set S by $n(S)$. If the sample space S is divided into mutually exclusive and equally likely outcomes, we define the *probability* of an event A by

$$p(A) = \frac{n(A)}{n(S)}. \tag{1.5}$$

In the example above, the sample space S was divided into four mutually exclusive and equally likely outcomes HH, HT, TH, and TT. If A is the event that Tom wins a dollar and $p(A)$ is its probability then $n(A) = 3$ and $n(S) = 4$. Hence $p(A) = 3/4$. Similarly if B is the event that Tom loses three dollars then $n(B) = 1$ and $p(B) = 1/4$.

The definition of probability given above can be easily generalized to allow outcomes that are not equally likely, but for simplicity we will assume equally likely outcomes throughout. From this definition $0 \leqslant p(A) \leqslant 1$ for any event A and 0 and 1 are the probabilities of impossible and certain events, respectively.

Example 24. If a pair of dice are rolled, there are 36 possible outcomes, as shown in Figure 1.1. Since it is clear that these outcomes are equally likely, the definition of probability can now be applied to answer many questions:

(1) What is the probability of rolling (a total of) 2? Since there is only one outcome with such a total, the answer is 1/36.

(2) What is the probability of rolling 7? There are six outcomes that total 7, so we get 6/36 or 1/6.

(3) What is the probability of rolling an even number? There are eighteen outcomes that show an even total, so the answer is 18/36 or 1/2.

In most problems the sample space is too large to allow a complete listing

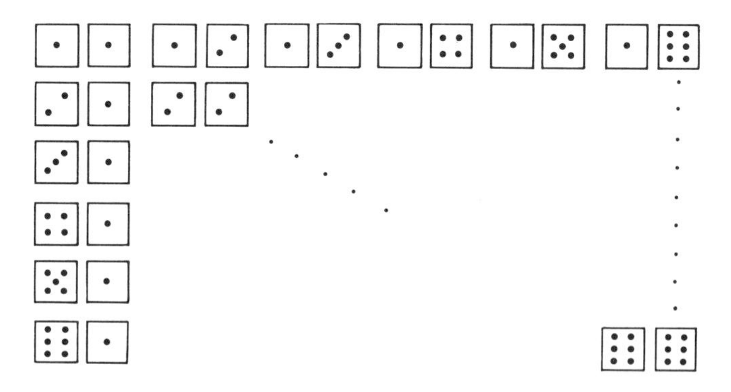

Figure 1.1

of elements, as in Example 24, so the counting techniques we have developed become essential. The next two examples illustrate this.

Example 25. Two cards are selected from a standard deck. What is the probability that they will both be aces? We need to count two things: the sample space and the number of outcomes that consist of two aces. Clearly we are making unordered selections without repetition, so we have $\binom{52}{2}$ outcomes in the sample space and $\binom{4}{2}$ successful outcomes. The probability of selecting two aces is therefore $\binom{4}{2}/\binom{52}{2} = 6/1326 = 1/221$.

Example 26. What is the probability that a number selected at random between 100 and 999 (inclusive) will not contain a 3? Here we must count the numbers between 100 and 999—there are $999 - 100 + 1 = 900$ of them. How many contain no 3? There are eight ways to choose the leading digit and nine ways to choose the second and the third, yielding a total of $8 \cdot 9 \cdot 9 = 648$ such numbers, by the multiplicative counting principle. Thus the probability of selecting such a number is $648/900 = 18/25$.

If an event has probability $p = n/s$, then the number $1 - p$ represents the probability it will *not* happen, since $1 - p = (s - n)/s$. This observation is used in the following example.

Example 27. What is the probability that a 5-card poker hand will contain an ace? It is easy to see that there are $\binom{52}{5}$ poker hands, but the number which contain an ace is harder to count. The number which contain *no* aces however, is $\binom{48}{5}$. Thus the probability of at least one ace is $1 - \{\binom{48}{5}/\binom{52}{5}\} \cong .341$.

EXERCISES

Exercises 1–4 each describe an experiment. For each experiment select a sample space in which the outcomes are equally likely. Then, for each space, (a) list two sample outcomes and (b) specify the total number of outcomes.

1. Three cards are selected from a standard deck and lined up on the table.

2. A committee of five people is selected randomly from a group of ten men (1 through 10) and ten women (a through j).

3. Joe, Mike, and Bill are each dealt a 5-card poker hand from a standard deck.

4. A 7-digit telephone number beginning with 484, 482, 491, 493, or 221 is chosen at random.

Exercises 5–8 describe events in the experiments of exercises 1–4. In each case, find the probability of the event described.

5. (a) The cards selected in exercise 1 are all red.
(b) No aces are selected.

6. The committee selected in exercise 2 contains more women than men.

7. (a) The three hands in exercise 3 contain only face cards (jacks, queens, and kings)
(b) The three hands are all flushes (i.e., each contains cards of only a single suit).

8. (a) In exercise 4, the number selected contains no 3's.
(b) The number contains exactly two 5's.

9. If a 5-digit number is chosen at random what is the probability it will contain at least one 7? That it will contain (at least) four 7's?

10. Joe went to the store to buy fruit, but he couldn't make up his mind about which of the five kinds he should buy. He asked the storekeeper to select four pieces for him, randomly. If the storekeeper does as he has been asked, what is the probability that Joe will get four fruits that are all the same kind?

11. If six people are in a queue at a movie theater, what is the probability that they are standing in alphabetical order?

1.6. COMPOUND EVENTS AND EXPECTATIONS

Probabilities can be added, subtracted, and multiplied, and in certain instances the resulting fractions will mean something. If events E_1 and E_2 have probabilities $p(E_1)$ and $p(E_2)$ and are disjoint, then the probability of the event $E_1 \cup E_2$ is $p(E_1) + p(E_2)$. In other words, if E_1 and E_2 cannot both occur, then the probability that one or the other of them will occur is the sum of their separate probabilities. This follows directly from the definition of probability. A more general result, stated in terms of the event $E_1 \cap E_2$ (E_1 and E_2 both occur) is the following.

Theorem 1.6. For any two events E_1 and E_2,

$$p(E_1 \cup E_2) = p(E_1) + p(E_2) - p(E_1 \cap E_2). \tag{1.6}$$

Proof. This follows immediately from the set-theoretic result that for any two sets S and T, $n(S \cup T) = n(S) + n(T) - n(S \cap T)$.

Example 28. What is the probability that a card drawn at random from a standard deck will be a spade or heart? The probability of drawing a spade is $13/52 = 1/4$, and the probability of drawing a heart is the same. Since a single card cannot be both a spade and a heart, $p(E_1 \cap E_2) = 0$, so the two probabilities are added, yielding 1/2 as the probability of a spade or a heart being drawn. This agrees with the number 26/52 obtained directly by counting the spades and hearts together.

Example 29. What is the probability that a card drawn at random from a standard deck will be a spade or an ace? The probability of drawing a spade is 13/52, the probability of an ace is 4/52, but it does *not* follow that the probability we seek is $13/52 + 4/52 = 17/52$, because the events "spade" and "ace" are not disjoint. The actual probability is 16/52, since the probability of drawing the ace of spades is 1/52, and we can apply formula (1.6).

Sometimes the probability $p(E_1 \cap E_2)$ that E_1 and E_2 will both occur is $p(E_1) \cdot p(E_2)$. To explain exactly when that is, we need a definition. The *conditional probability* $p(E_2|E_1)$ is the probability that E_2 will occur, given that E_1 occurs. We claim that

$$p(E_2|E_1) = n(E_1 \cap E_2)/n(E_1). \tag{1.7}$$

This is true since our sample space contains only the outcomes in E_1 (since we assume E_1 occurs) and $n(E_1 \cap E_2)$ counts the outcomes in E_2 that lie within E_1.

Example 30. Two dice are rolled, and one goes under the table. If the red die (which is showing) is a 3, what is the probability that the total on the two dice is 8? Since we know the red die is 3, the sample space contains not 36 outcomes, but only 6. Of those 6 one has a total of 8. Thus the probability is 1/6 that the total is 8, given that the red die is a 3.

We can now state just how $p(E_1 \cap E_2)$ is found.

Theorem 1.7. For any two events E_1 and E_2,

$$p(E_1 \cap E_2) = p(E_1)p(E_2|E_1). \tag{1.8}$$

Proof. If S is the sample space, the left-hand side is just $n(E_1 \cap E_2)/n(S)$. The two terms on the right are $n(E_1)/n(S)$ and $n(E_1 \cap E_2)/(E_1)$, respectively, which proves the equation.

If $p(E_2|E_1) = p(E_2)$ we say that E_1 and E_2 are *independent* events. Intuitively, this means that the two events do not have anything to do with each other. Care must be taken, however, in claiming that the two events are or are not independent. Note that if events are independent, (1.8) says that the probability they will both occur is the product of the individual probabilities. Of course, this extends to more than two independent events as well.

Example 31. If one card is drawn from each of two decks, what is the probability that both cards are aces? Since the two draws are clearly independent, the answer is $(4/52)^2 = (1/13)^2 = 1/169$.

Example 32. If two cards are drawn from a single deck, what is the probability that both cards are aces? The two draws are not independent in this case. The obvious way to find the probability is to simply count possibilities. The number of ways of drawing two cards is $\binom{52}{2}$ and the number of ways of drawing two aces is $\binom{4}{2}$. Thus the probability of aces is $\binom{4}{2}/\binom{52}{2} = 1/221$. Alternatively, we can use formula (1.8) with E_1 being an ace on the first draw and E_2 an ace on the second. Clearly $p(E_1) = 4/52$ and $p(E_2|E_1) = 3/51$. Thus $p(E_1 \cap E_2) = (4/52)(3/51) = 1/221$, just as before.

Probability cannot tell us what will happen on any particular trial of a game or experiment, but is useful only as a predictor of long-term trends. A concept known as the expectation, or expected value, of a game is based on this knowledge of long-term behavior and can help us decide whether or not a particular game is to our advantage.

Consider the following trivial game. We pay one dollar and then flip a coin. If the coin is heads, we win two dollars; if it is tails we get nothing. Is this a good game to play? We can win or lose a dollar each time, depending on whether we are lucky or unlucky. It is fairly clear that we should come out about even in the long run, since heads or tails are equally probable. How can this be formalized?

In a game involving payoffs (cash or otherwise), the outcomes can be partitioned into n events $E_1, E_2, \ldots, E_n$, where all of the outcomes in event E_i have the same payoff, which we denote by $P(E_i)$. The *expectation* of the game is the sum

$$e = p(E_1)\,P(E_1) + p(E_2)P(E_2) + \cdots + p(E_n)P(E_n). \tag{1.9}$$

The expectation is then the average amount one might expect to win or lose per game, in the long run. Looking at our coin example, the payoff for a head is one dollar (we pay a dollar, and get back two), and the payoff for a tail is minus one dollar. Thus our expectation is $(1/2)(1) + (1/2)(-1) = 0$, as we expected. Thus, in the long run we should come out about even. Note that the expectation is not necessarily what we will get on any particular trial, but how we can expect to fare in the long run.

Example 33. 10,000 people enter a mail sweepstakes, in which the prize is a \$500 color television. What is the expectation of each entrant? There are two events to be considered: winning ($p = 1/10{,}000$) and losing ($p = 9999/10{,}000$). The prize is \$500, the cost of entering is \$0.20 for postage. Thus $e = (1/10{,}000)(500 - 0.20) + (9999/10{,}000)(-0.20) = -\$0.15$. Thus there is a negative expectation, and if you played this game often you would lose an average of 15 cents per game in the long run.

Example 34. If six holes are each filled with a peg of one of seven colors (repetitions allowed), and this is then done again, randomly, what is the expected number of holes filled the same way both times? (Those people who play the game Mastermind will recognize this situation, but a knowledge of the game is not needed for the problem.) The counting solution is as follows. There are 7^6 ways of filling the spaces a second time. We divide these 7^6 outcomes into events according to the number of matches with the original placement of pegs. If event E_i contains those placements that give $6 - i$ matches, for $i = 0$ to 6, there are $\binom{6}{i} \cdot 6^i$ placements in event i. This is so because we can chose the i spaces that fail to match in $\binom{6}{i}$ ways, and each can then be filled in six ways, while the other $6 - i$ places can be filled in only one way. Thus the expected number of matches is

$$\sum_{i=0}^{6} \frac{\binom{6}{i} \cdot 6^i}{7^6} (6 - i). \tag{1.10}$$

A bit of arithmetic shows that this is 6/7.

The simple answer suggests that we could probably do the preceding example an easier way.

In fact the observation that simplifies the problem is that we can concentrate attention on a single hole. In hole number 1, we have a match with probability 1/7 and no match with probability 6/7. Thus the expected number of matches in hole 1 is $1/7 \cdot 1 + 6/7 \cdot 0 = 1/7$. This is also the expected number of matches in any of the other holes and adding over the six holes we obtain 6/7, as before.

This process of addition is justified not merely because it gives the right answer (although that is nice), but because expectations are, in general, additive. If you play several games, and expect to win $1 on the first, lose $2 on the second, and win $3 on the third, it is fairly clear that your total expectation is $2. Thus expectations are nicer than probabilities. If your probability of winning the three games are, respectively, 1/2, 1/8, and 1/2, it is clear you do not want to say that your probability of winning one of them is 9/8.

EXERCISES

1. If a single card is drawn from a standard deck, what is the probability that it is red or a face card?

2. If two dice are rolled, what is the probability that at least one will show a 4? What if three dice are rolled?

3. What is the probability that in eight tosses of a fair coin there will be exactly four heads? At least four heads?

4. John has a penny, a nickel, a dime, and a quarter to distribute among six friends. If he distributes them randomly, what is the probability that some friend will get more than one coin? Same question, if John has four pennies.

5. If two dice show a total of 8, what is the probability that one of the dice shows a 3? A 4?

6. If two dice are thrown, are the following events independent? E_1: a total of 7. E_2: both dice show the same number or one of the dice is a 1 and the other is even.

7. You and your opponent are each dealt four cards from a standard deck. If you have no aces, what is the probability that she has at least one ace?

8. In a hurry at the airport, you grab a bill from your wallet to pay the cab driver. If you had six tens and four fives in your wallet, what is the cab driver's expected take from the deal?

9. You play a card game in which you pay $10 to play each round. Then you draw a card from the deck. If it is an ace you get $50, if it is a king you get $20, and if it is a queen or jack you get $15. Otherwise you get nothing. Is this a good game to play?

10. Among thirty people chosen at random, what is the probability that at least two of them have the same birthday? [*Hint:* find the probability that all the birthdays are different.]

1.7. REMARKS

Beginning with two simple counting principles, we have developed an extensive collection of techniques to solve problems of choice and arrangement. The application of those techniques is anything but simple, requiring careful and clever reasoning. In the next two chapters we will develop several more sophisticated counting procedures.

Some of the methods and results in this chapter have been known for a very long time. The process of uniting them into a cohesive theory was initiated by Pascal and Fermat in 16th-century France. They were motivated by a desire to accurately analyze certain games of chance. The book by David [1] contains a wealth of information about the history of counting and its strong link to gambling. The move to academic respectability has been slow. In fact, the first text on counting methods was published in 1897, and it is still available in reprint form [2].

Modern accounts of elementary counting techniques include a nice little book by Niven [3] and a translation, from the Russian, of a book by Vilenkin [4] containing almost 500 solved problems. Most introductory combinatorics texts begin with counting methods. Three of the best choices are Liu [5], Cohen [6], and Tucker [7]. The recent book by Packel [8] explains probability for the gambler.

[1] F. N. David, *Games, Gods, and Gambling*, Hafner, New York, 1962.

[2] W. Whitworth, *Choice and Chance*, Hafner Press, New York, 1965.

[3] I. Niven, *Mathematics of Choice*, MAA, Washington, 1965.

[4] N. Ya. Vilenkin, *Combinatorics*, Academic Press, New York, 1971.

[5] C. L. Liu, *Introduction to Combinatorial Mathematics*, McGraw-Hill, New York, 1968.

[6] D. Cohen, *Basic Techniques of Combinatorial Theory*, Wiley, New York, 1978.

[7] A. Tucker, *Applied Combinations*, Wiley, New York, 1980.

[8] E. Packel, *The Mathematics of Games and Gambling*, MAA, New York, 1981.

CHAPTER 2

The Binomial Theorem

In this chapter we continue our development of counting techniques by making a connection between the number of subsets of size k in an n-set and the coefficient of x^k in a polynomial of degree n. Polynomials that have certain numbers as coefficients occur frequently in combinatorial enumeration, because when we deal with polynomials a great deal of information can be carried by a single expression and we have the tools of algebra and analysis at our disposal. In fact we are sometimes concerned with infinite strings of numbers, and this leads to the use of infinite series.

2.1. THE BINOMIAL THEOREM

We observed in the last chapter that $\binom{n}{r}$ is the number of r-subsets of an n-set. In fact such numbers turn up in some rather unexpected places, and the following theorem is an outstanding example of their importance.

Theorem 2.1. (The binomial theorem). If n is a non-negative integer then

$$(1 + x)^n = \sum_{r=0}^{n} \binom{n}{r} x^r. \tag{2.1}$$

Proof. This can be proved by induction, but we choose a more intuitive approach. Note that $(1 + x)^n$ means $(1 + x)(1 + x) \cdots (1 + x)$, where the product extends to n terms. What does such an expression mean? Let us reduce to the case $n = 3$ and place subscripts on the x's so we can keep track of them. The product then becomes $(1 + x_1)(1 + x_2)(1 + x_3)$, which is $1 \cdot 1 \cdot 1 + x_1 \cdot 1 \cdot 1 + 1 \cdot x_2 \cdot 1 + 1 \cdot 1 \cdot x_3 + x_1 \cdot x_2 \cdot 1 + x_1 \cdot 1 \cdot x_3 + 1 \cdot x_2$

$\cdot x_3 + x_1 \cdot x_2 \cdot x_3$, reducing to $1 + 3x + 3x^2 + 3x^3$ when the subscripts are removed. To obtain this product we have just multiplied every possible choice of factors together and added all of the results. The coefficient 3 appears because there are three ways we can choose a single x and two 1's to multiply together. In the general case, the coefficient of x^r is the number of ways we can choose r factors that contribute x (leaving $n - r$, which contribute 1) to obtain a product x^r. Since there are $\binom{n}{r}$ ways to make such a choice of factors, the coefficient of x^r in the product must be $\binom{n}{r}$, as claimed.

Because of this theorem, the numbers $\binom{n}{r}$ are often called *binomial coefficients*. The theorem can be used to prove many identities on these numbers, by letting x take on various values or using algebraic manipulation.

Example 1. Let $x = 1$ in equation (2.1), to obtain the equation $2^n = \binom{n}{0} + \binom{n}{1} + \cdots + \binom{n}{n}$. This can be viewed as counting the subsets of an n-set two ways: The total number is 2^n and there are $\binom{n}{i}$ of size i, for each i from 0 to n.

Example 2. Let $x = -1$ to obtain $0 = \binom{n}{0} - \binom{n}{1} + \cdots + (-1)^n\binom{n}{n}$. This implies that $\binom{n}{0} + \binom{n}{2} + \cdots = \binom{n}{1} + \binom{n}{3} + \cdots$, so the number of even subsets of an n-set equals the number of odd subsets.

Example 3. Integrating equation (2.1), we obtain

$$\frac{1}{n+1}(1 + x)^{n+1} = \binom{n}{0}x + \frac{1}{2}\binom{n}{1}x^2 + \cdots + \frac{1}{n+1}\binom{n}{n}x^{n+1} + c.$$

Clearly c must be $1/(n + 1)$ to make this work. Setting $x = 1$ we find

$$\frac{1}{n+1}2^{n+1} = \binom{n}{0} + \frac{1}{2}\binom{n}{1} + \cdots + \left(\frac{1}{n+1}\right)\binom{n}{n} + \frac{1}{n+1}$$

so that

$$\binom{n}{0} + \frac{1}{2}\binom{n}{1} + \cdots + \frac{1}{n+1}\binom{n}{n} = \frac{1}{n+1}(2^{n+1} - 1). \qquad (2.2)$$

Example 4. Differentiating equation (2.1) and setting $x = -1$, we obtain the identity $0 = \binom{n}{1} - 2\binom{n}{2} + \cdots + (-1)^{n-1}n\binom{n}{n}$. If we set x equal to 1 instead, we have $n2^{n-1} = \binom{n}{1} + 2\binom{n}{2} + \cdots + n\binom{n}{n}$. These equations can be used to derive several others. For example, if we add them for n even we obtain $n2^{n-2} = \binom{n}{1} + 3\binom{n}{3} + \cdots + (n - 1)\binom{n}{n-1}$.

These examples illustrate the convenience gained by thinking of the numbers $\binom{n}{r}$ as binomial coefficients. Although each of those results can be derived using the original factorial definition of the numbers, a direct arithmetic attack is extremely laborious. More identities will be derived in the next section, using various methods.

Of course, binomials are not always of the form $(1 + x)^n$, but may be more general. The following corollary of Theorem 2.1 shows how to handle the more general case.

Corollary 2.1. If n is a non-negative integer then

$$(x + y)^n = \sum_{r=0}^{n} \binom{n}{r} x^{n-r} y^r.$$

Proof. Note that

$$(x + y)^n = \left[x\left(1 + \frac{y}{x}\right)\right]^n = x^n\left(1 + \frac{y}{x}\right)^n.$$

By the binomial theorem, we know how to expand $(1 + y/x)^n$, so this is

$$x^n \sum_{r=0}^{n} \binom{n}{r}\left(\frac{y}{x}\right)^r = x^n \sum_{r=0}^{n} \binom{n}{r} y^r x^{-r}.$$

Bringing the x^n inside the sum, we obtain

$$\sum_{r=0}^{n} \binom{n}{r} y^r x^{-r} x^n = \sum_{r=0}^{n} \binom{n}{r} x^{n-r} y^r$$

as desired.

Example 5. What is the coefficient of $x^{11}y^4$ in the expansion of $(2x + y)^{15}$? The term in question has the form $\binom{15}{4}(2x)^{11}y^4$, so the coefficient of $x^{11}y^4$ is $\binom{15}{4}2^{11}$, or 2,795,520.

EXERCISES

1. Give an alternate proof of the binomial theorem in the following way. Let $(1 + x)^n = a_0 + a_1x + \cdots + a_nx^n$. Set $x = 0$ to find a_0. Differentiate and set $x = 0$ to find a_1. Continue in this way to find all of the a_n and thus prove the theorem.

2. In the expansion of $(x + y)^n$ there is a term of the form Ax^5y^m, A a constant. What is A and what is m?

3. What is the coefficient of x^7 in $(2 + 3x)^{10}$?

4. (a) What is the coefficient of x^4 in the expansion of $(1 + x + x^2)(1 + x)^5$?
(b) What is the coefficient of x^{12} in the expansion of $(1 + x + x^{-1})(1 + x)^{26}$?

5. There are six boxes lined up in a row. The first contains two balls, the others each contain one. In how many ways can you select four balls from the boxes? [*Hint:* use exercise 4(a).]

6. Find the binomial sums using identities from this section.
(a) $\binom{n}{0} + \binom{n}{2} + \binom{n}{4} + \binom{n}{6} + \cdots$
(b) $\binom{n}{1} + \binom{n}{3} + \binom{n}{5} + \binom{n}{7} + \cdots$
(c) $\binom{n}{0} + 2\binom{n}{1} + \binom{n}{2} + 2\binom{n}{3} + \binom{n}{4} + 2\binom{n}{5} + \cdots$
(d) $\binom{n}{0} + 3\binom{n}{1} + 5\binom{n}{2} + 7\binom{n}{3} + \cdots$

7. Find the following binomial sums by manipulation of equation (2.1) and setting x to some particular value.

(a) $\sum_{r=0}^{n} 2^r \binom{n}{r}$

(b) $\sum_{r=2}^{n} r(r-1)\binom{n}{r}$

(c) $\sum_{r=2}^{n} \binom{r}{2}\binom{n}{r}$

(d) $\sum_{r=0}^{n} (2r+1)\binom{n}{r}$

(e) $\sum_{r=1}^{n/2} 2r\binom{n}{2r}$ (n even)

(f) $\sum_{r=1}^{(n+1)/2} (2r-1)\binom{n}{2r-1}$ (n odd)

2.2. BINOMIAL IDENTITIES

The equations derived in the last section are just the beginning of a list of hundreds of such identities that appear in the literature. We will derive a few more in this section.

Probably the most useful of all binomial identities is

$$\binom{n}{r} = \binom{n-1}{r} + \binom{n-1}{r-1}. \tag{2.3}$$

This can be proved easily in at least three ways. Writing each coefficient in factorial form and combining the two fractions on the right-hand side over a common denominator is the most obvious approach. Thus,

$$\begin{aligned}\binom{n-1}{r} + \binom{n-1}{r-1} &= \frac{(n-1)!}{r!(n-r-1)!} + \frac{(n-1)!}{(r-1)!(n-r)!}\\ &= \frac{(n-1)!}{(r-1)!(n-r-1)!}\left[\frac{1}{r} + \frac{1}{n-r}\right]\\ &= \frac{(n-1)!}{(r-1)!(n-r-1)!}\cdot\frac{n}{r(n-r)}\\ &= \frac{n!}{r!(n-r)!}\\ &= \binom{n}{r}.\end{aligned}$$

A second method involves thinking of $\binom{n}{r}$ as the number of ways of choosing r objects from an n-set. If one object is distinguished, then there are clearly $\binom{n-1}{r}$ choices of r objects that do not include the distinguished one and $\binom{n-1}{r-1}$ choices that do include it, yielding the identity. Our final proof uses the equation $(1 + x)^n = (1 + x)(1 + x)^{n-1}$. The coefficient of x^r on the left-hand side is $\binom{n}{r}$, and an x^r can be obtained on the right-hand side by multiplying 1 by x^r or by multiplying x by x^{r-1}. The coefficient of x^r in $(1 + x)^{n-1}$ is $\binom{n-1}{r}$, and the coefficient of x^{r-1} is $\binom{n-1}{r-1}$, yielding the identity once again.

The method of equating coefficients is also used to prove

$$\binom{n+m}{r} = \binom{n}{0}\binom{m}{r} + \binom{n}{1}\binom{m}{r-1} + \cdots + \binom{n}{n}\binom{m}{r-n} \quad (r \geqslant n). \tag{2.4}$$

The binomial theorem gives us

$$(1 + x)^n = \binom{n}{0} + \binom{n}{1}x + \cdots + \binom{n}{n}x^n$$

$$(1 + x)^m = \binom{m}{0} + \binom{m}{1}x + \cdots + \binom{m}{m}x^m.$$

Multiplying the left- and right-hand sides of these equations, we obtain

$$(1 + x)^{n+m} = \left[\binom{n}{0} + \binom{n}{1}x + \cdots + \binom{n}{n}x^n\right]\left[\binom{m}{0} + \binom{m}{1}x + \cdots + \binom{m}{m}x^m\right].$$

Since the coefficients of x^r on the left- and right-hand sides of this equation must be equal, we have proved equation (2.4).

Notice that both (2.3) and (2.4) express a binomial coefficient in terms of smaller ones. In particular, (2.3) reduces a problem on n objects to two problems on $n - 1$ objects. This makes it useful in inductive proofs and also in computations. If we define $\binom{n}{r} = 0$ for $r > n$ and recall that $\binom{n}{0} = 1$ for all $n > 0$, we have the numbers $\binom{n}{r}$ for n or r zero as shown in Table 2.1(a). Applying (2.3) as indicated by the circles and arrows, we can fill in successive rows, obtaining Table 2.1(b). The nonzero entries of Table 2.1(b) form *Pascal's Triangle*. Note that the nonzero entries in row n are the coefficients in the expansion of $(1 + x)^n$.

Examination of the numbers suggests various identities. The symmetry in each row is explained by the identity

$$\binom{n}{r} = \binom{n}{n - r}, \tag{2.5}$$

Table 2.1. The numbers $\binom{n}{r}$

(a)

n \ r	0	1	2	3	4	5	...
0	1	0	0	0	0	0	...
1	1						
2	1						
3	1						
4	1						
5	1						
.	.						
.	.						

(b)

n \ r	0	1	2	3	4	5	...
0	(1) +	(0)	0	0	0	0	...
1	1 ↘	(1)	0	0	0	0	...
2	1	2	1	0	0	0	...
3	1	3	3	1	0	0	...
4	1	4	(6) +	(4)	1	0	...
5	1	5	10 ↘	(10)	5	1	...
.	.	.	.	.	.	.	...
.	.	.	.	.	.	.	...

which was mentioned in exercise 2 in Section 1.3. Summing the first n entries in a column we obtain the $(n + 1)$st term in the next column. In symbols, this is

$$\binom{0}{r} + \binom{1}{r} + \binom{2}{r} + \cdots + \binom{n}{r} = \binom{n+1}{r+1}. \tag{2.6a}$$

It is easy to prove equation (2.6a) inductively. Note that $\binom{0}{r} = \binom{1}{r+1}$, since both sides are either 0 or both sides are 1, depending on whether r is larger than or equal to 0. Suppose then we have shown that $\binom{0}{r} + \binom{1}{r} + \cdots + \binom{n-1}{r} = \binom{n}{r+1}$. Adding $\binom{n}{r}$ to both sides, and using (2.3) on the right, we have (2.6a).

Since $\binom{i}{r} = 0$ for $i < r$, (2.6a) can be reduced to

$$\binom{r}{r} + \binom{r+1}{r} + \cdots + \binom{n}{r} = \binom{n+1}{r+1}. \tag{2.6b}$$

Several special cases of equation (2.6b) may be familiar to you. If $r = 0$ it takes the form $1 + 1 + 1 + \cdots + 1 = n + 1$, which is not particularly surprising! For $r = 1$ we get the more interesting case

$$1 + 2 + 3 + \cdots + n = \frac{n(n+1)}{2}. \tag{2.7}$$

Finally, for $r = 2$ we have

$$1 + 3 + 6 + 10 + \cdots + \frac{n(n+1)}{2} = \frac{(n+2)(n+1)n}{6}. \tag{2.8}$$

The numbers obtained as sums of consecutive integers in (2.7) are called *triangular numbers* because such a number of dots can be arranged into an equilateral triangle, as shown in Figure 2.1. Triangular numbers are just the binomial coefficients $\binom{n}{2}$. Note that the numbers being added in equation

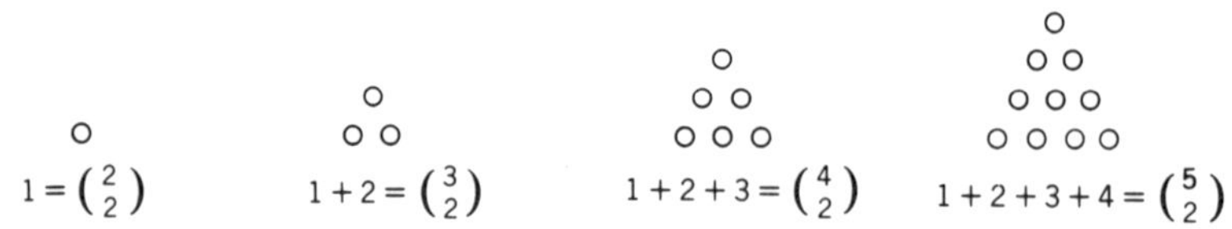

$1 = \binom{2}{2}$ $1+2 = \binom{3}{2}$ $1+2+3 = \binom{4}{2}$ $1+2+3+4 = \binom{5}{2}$

Figure 2.1. Triangular Numbers

(2.8) are triangular since they are $\binom{i}{2}$ for some i. The adding together of such numbers can be viewed geometrically as a stacking of the corresponding triangles. Thus the numbers $\binom{n}{3}$ obtained as the sum of triangular numbers in equation (2.8) are called *pyramidal*. Unfortunately, this nice geometric interpretation of $\binom{n}{r}$ can be carried on for $n > 3$ only by those who have good geometric intuitions in spaces of four or more dimensions!

The display of the numbers $\binom{n}{r}$ in Table 2.1 also suggests an application to paths. If you begin at $\binom{0}{0}$ and proceed to other numbers by steps that are either directly downward (D) or diagonally downward to the right (R), the number reached will equal the number of ways of reaching that number. (For example, $\binom{2}{1} = 2$ is reached by the sequence of steps DR and RD.) To see why this is so, note that the number $\binom{n}{r}$ is reached in exactly n steps r of which are R. Furthermore, any choice of n steps, r of which are R, must reach $\binom{n}{r}$. But there are $\binom{n}{r}$ such sequences of steps. Another proof uses (2.3) and induction to n. Every path to $\binom{n}{r}$ is just a path to $\binom{n-1}{r}$ plus a step D, or a path to $\binom{n-1}{r-1}$ plus a step R. Assuming there are $\binom{n-1}{r}$ paths to $(n - 1, r)$ and $\binom{n-1}{r-1}$ paths to $(n - 1, r - 1)$, we find there are $\binom{n-1}{r} + \binom{n-1}{r-1} = \binom{n}{r}$ paths to (n, r).

EXERCISES

1. Prove the following identities by writing out both sides in factorials and simplifying.

(a) $\frac{n+1}{r+1}\binom{n}{r} = \binom{n+1}{r+1}$.

(b) $\binom{n}{m}\binom{m}{r} = \binom{n}{r}\binom{n-r}{m-r}$

2. Prove the following identities, using induction.

(a) $\binom{n}{0} + \binom{n+1}{1} + \binom{n+2}{2} + \cdots + \binom{n+r}{r} = \binom{n+r+1}{r}$

(b) $\binom{n}{0} - \binom{n}{1} + \binom{n}{2} + \cdots + (-1)^r\binom{n}{r} = (-1)^r\binom{n-1}{r}$

3. Prove the following identities by equating coefficients in two different but equal expressions, or by using some other combinatorial argument.

(a) $\binom{n}{0} + \binom{n+1}{1} + \binom{n+2}{2} + \cdots + \binom{n+r}{r} = \binom{n+r+1}{r}$

(b) $\sum_{r=0}^{n} \binom{n}{r}\binom{m}{k+r} = \binom{n+m}{n+k}$

(c) $\binom{n}{1} + 2\binom{n}{2} + \cdots + n\binom{n}{n} = n2^{n-1}$

(d) $\binom{n}{0}^2 + \binom{n}{1}^2 + \binom{n}{2}^2 + \cdots + \binom{n}{n}^2 = \binom{2n}{n}$

4. By counting paths in Pascal's Triangle, prove the following identity.

$$\binom{n}{0} + \binom{n+1}{1} + \binom{n+2}{2} + \cdots + \binom{n+r}{r} = \binom{n+r+1}{r}.$$

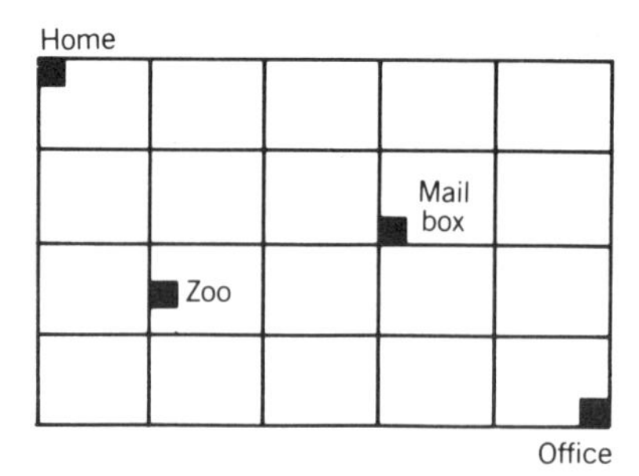

5. Mary walks from her home to her office along city streets, as shown above. If she always walks a shortest way
   (a) In how many different ways can she walk to her office?
   (b) How many of those different ways take her past the mailbox?
   (c) If she chooses her route at random, what is the probability she will walk past the zoo?

6. Consider the coefficients in the expansion of $(1 + x + x^2)^n$. If we call the coefficient of x^r in this expansion nBr, the numbers nBr can be arranged in a triangular array with first row 1, second row 1 1 1, third row 1 2 3 2 1, and so on.
   (a) Find a rule that gives the entries in each row in terms of the entries in the previous row, and state it as an identity on nBr's.
   (b) Show $nB0 + nB1 + nB2 + \cdots + nB2n = 3^n$.
   (c) Show $nB0 - nB1 + nB2 - \cdots + nB2n = 1$.

7. Exactly 2^{1000} balls are fed into the top of the sorter illustrated below. We have shown the beginning of the machine; it actually contains 1001

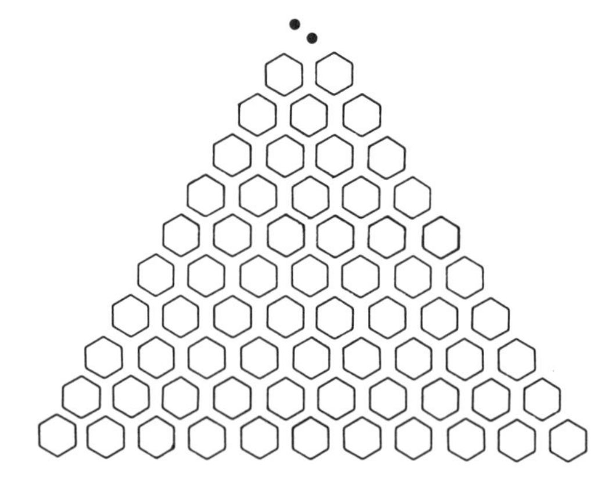

rows of hexagons. If we assume a perfect sort, so that whenever two balls reach a junction one goes left and the other right, how many balls wind up in the left-most compartment at the bottom? How many wind up in compartment k (from the left)?

2.3. MULTINOMIAL COEFFICIENTS AND NEGATIVE BINOMIAL COEFFICIENTS

If you understood the proof of the binomial theorem, you should have no trouble proving the generalization to multinomials. The essential fact is that $\binom{n}{r_0, r_1, \ldots, r_k}$ counts the number of ways of choosing n objects, including r_i of type i for each i ($i = 0, 1, 2, \ldots, k$).

Theorem 2.2. (The multinomial theorem) If n is a non-negative integer then

$$(1 + x_1 + x_2 + \cdots + x_k)^n = \sum \binom{n}{r_0, r_1, \ldots, r_k} 1^{r_0} x_1^{r_1} x_2^{r_2} \cdots x_k^{r_k} \quad (2.9)$$

where the sum is over all choices of the r_i's as non-negative integers that sum to n.

The applications of Theorem 2.2 are similar to those of the binomial theorem, although more complicated and less frequently encountered. We present only three examples.

Example 6. Setting $x_1 = x_2 = \cdots = x_k = 1$ in (2.9) we find the identity

$$(k + 1)^n = \sum \binom{n}{r_0, r_1, \ldots, r_k}. \quad (2.10)$$

Example 7. Factoring $(1 + x_1 + x_2 + \cdots + x_k)^n$ as $(1 + x_1 + x_2 + \cdots + x_k)^{n-1} \cdot (1 + x_1 + x_2 + \cdots + x_k)$ and equating coefficients of the terms $x_1^{r_1} x_2^{r_2} \cdots x_k^{r_k}$, we find

$$\binom{n}{r_0, r_1, \ldots, r_k} = \sum \binom{n - 1}{r_0, r_1, \ldots, r_i - 1, \ldots, r_k}, \quad (2.11)$$

where the sum extends over those i for which $r_i \neq 0$. Thus, for example,

$$\binom{14}{2, 3, 4, 5} = \binom{13}{1, 3, 4, 5} + \binom{13}{2, 2, 4, 5} + \binom{13}{2, 3, 3, 5} + \binom{13}{2, 3, 4, 4}.$$

Note the special case $\binom{n}{r_1, r_2} = \binom{n-1}{r_1 - 1, r_2} + \binom{n-1}{r_1, r_2 - 1}$, which is more familiar to us as $\binom{n}{r} = \binom{n-1}{r-1} + \binom{n-1}{r}$.

Example 8. For the case $k = 2$, Example 7 yields

$$\binom{n}{r_0, r_1, r_2} = \binom{n-1}{r_0 - 1, r_1, r_2} + \binom{n-1}{r_0, r_1 - 1, r_2} + \binom{n-1}{r_0, r_1, r_2 - 1}.$$

An array of values can be made in the first octant of 3-space, with the number $\binom{r_0 + r_1 + r_2}{r_0, r_1, r_2}$ at (r_0, r_1, r_2), which is the 3-dimensional analogue of Pascal's Triangle. The number of paths from (0, 0, 0) to (r_0, r_1, r_2), which always proceed from one lattice-point to an adjacent larger one, will be $\binom{r_0+r_1+r_2}{r_0, r_1, r_2}$

Another way to generalize the binomial theorem is to allow negative integral powers. In order to cope with expansions of negative powers, we must use the theory of infinite series. Recall that the geometric series $a + ar + ar^2 + \cdots + ar^n + \cdots$, with initial term a and ratio r, has sum $a/(1 - r)$. This is easy to see, since if $S = a + ar + ar^2 + \cdots + ar^n + \cdots$ then $Sr = ar + ar^2 + \cdots + ar^n + \cdots$. Subtracting, we find $S - Sr = S(1 - r) = a$, so that $S = a/(1 - r)$. The radius of convergence of this series is 1, so the sum converges if and only if $|r| < 1$. These basic facts about infinite series are keys to expanding expressions $(1 + x)^{-n}$.

Theorem 2.3. If n is a positive integer then

$$(1 + x)^{-n} = \sum_{j=0}^{\infty} (-1)^j \binom{n + j - 1}{j} x^j. \tag{2.12}$$

Proof. We first deal with the case $n = 1$. Note that $(1 - x)^{-1}$ is the sum of the series $1 + x + x^2 + \cdots + x^n + \cdots$. Replacing x by $-x$ in this expression we find

$$(1 + x)^{-1} = 1 - x + x^2 + \cdots + (-1)^k x^k + \cdots, \tag{2.13}$$

which is equation (2.12) with $n = 1$.

The formula (2.12) for $n > 1$ is derived from (2.13) by repeated differentiation. It is easy to verify that the $(n - 1)$st derivative of $(1 + x)^{-1}$ is $(-1)^{n-1}(n - 1)!(1 + x)^{-n}$, while the $(n - 1)$st derivative of x^k is $k(k - 1)\ldots(k - n + 2)x^{k-n+1}$. Keeping these results in mind, we differentiate both sides of equation (2.13) $n - 1$ times, to obtain

$$(-1)^{n-1}(n-1)!(1+x)^{-n} = \sum_{k=n-1}^{\infty} (-1)^k k(k-1)\ldots(k-n+2)x^{k-n+1}. \quad (2.14)$$

We divide both sides by $(-1)^{n-1}(n-1)!$ and omit the first few terms of the sum (which are zero) to obtain

$$(1+x)^{-n} = \sum_{k=n-1}^{\infty} \frac{(-1)^{k-n+1}k(k-1)\ldots(k-n+2)}{(n-1)!x^{k-n+1}}. \quad (2.15)$$

Letting $j = k - n + 1$, we have $k = j + n - 1$, and (2.15) is equivalent to (2.12) as desired.

This binomial theorem for negative exponents is not just an exhibition of symbol-pushing, but has combinatorial significance. Note that the coefficient of x^j in the expansion of $(1+x)^{-n}$ is $(-1)^j$ times the number of j-selections from an n-set. Clearly, if we choose j objects from an n-set and allow repetition, the number j can be as large as we like. Thus an expression that has as its coefficients the number of selections, rather than the number of combinations, must necessarily be infinite rather than finite.

If some care is exercised, Table 2.1(b) can be extended to a table that contains all of the negative binomial coefficients:

$$\binom{-n}{j} = (-1)^j \binom{n+j-1}{j}. \quad (2.16)$$

If we suppose that we want $\binom{n}{r}$ and $\binom{-n}{r}$ to be 0 for $r < 0$, and wish to preserve our rule that every entry is the sum of two entries, those directly above, and above and to the left, then the entries $\binom{n}{r}$, $r < 0$, and $\binom{-1}{r}$ are determined as shown in Table 2.2(a). Continuing in this way, filling in one row at a time, we obtain the whole table, which begins as shown in Table 2.2(b). Notice that the coefficients $\binom{-n}{r}$ are never 0 for $r \geqslant 0$. It is not difficult to show that the numbers $\binom{-n}{r}$ satisfy the relationship

$$\binom{-n}{j} + \binom{-n}{j-1} = \binom{-n+1}{j} \quad (2.17)$$

In fact, we have

$$\binom{-n}{j} + \binom{-n}{j-1} = (-1)^j\binom{j+n-1}{j} + (-1)^{j-1}\binom{j+n-2}{j-1}$$

Table 2.2. More numbers $\binom{n}{r}$.

$n \backslash r$	...	−2	−1	0	1	2	3	4	5	..
n	...	.	.	.	.	.	.	.	.	..
.	...	.	.							
.	...	.	.							
.	...	.	.							
−3	...	0	0	1						
−2	...	0	0	1						
−1	...	0	0	1	−1	1	−1	1	−1	..
0	...	0	0	1	0	0	0	0	0	..
1	...	0	0	1	1	0	0	0	0	..
2	...	0	0	1	2	1	0	0	0	..
3	...	0	0	1	3	3	1	0	0	..
4	...	0	0	1	4	6	4	1	0	..
.	...	.	.	.	.	.	.	.	.	..
.	...	.	.	.	.	.	.	.	.	..
.	...	.	.	.	.	.	.	.	.	..

(a)

$n \backslash r$	...	−2	−1	0	1	2	3	4	5	..
.	...	.	.	.	.	.	.	.	.	..
.	...	.	.	.	.	.	.	.	.	..
.	...	.	.	.	.	.	.	.	.	..
−3	...	0	0	1	−3	6	−10	15	−21	..
−2	...	0	0	1	−2	3	−4	5	−6	..
−1	...	0	0	1	−1	1	−1	1	−1	..
0	...	0	0	1	0	0	0	0	0	..
1	...	0	0	1	1	0	0	0	0	..
2	...	0	0	1	2	1	0	0	0	..
3	...	0	0	1	3	3	1	0	0	..
4	...	0	0	1	4	6	4	1	0	..
.	...	.	.	.	.	.	.	.	.	..
.	...	.	.	.	.	.	.	.	.	..
.	...	.	.	.	.	.	.	.	.	..

(b)

$$= (-1)^j \frac{(j+n-1)!}{j!(n-1)!} + (-1)^{j-1} \frac{(j+n-2)!}{(j-1)!(n-1)!}$$

$$= (-1)^j \frac{(j+n-2)!}{(j-1)!(n-1)!} \left\{ \frac{j+n-1}{j} - 1 \right\}$$

$$= (-1)^j \frac{(j+n-2)!}{j!(n-1)!}$$

$$= (-1)^j \binom{j+n-2}{j}$$

$$= \binom{-n+1}{j}.$$

With the definition (2.16), we can now write the binomial theorem with negative integral exponent as

$$(1+x)^{-n} = \sum_{r=0}^{\infty} \binom{-n}{r} x^r. \qquad (2.18)$$

EXERCISES

1. What is the sum of the coefficients in the expansion of $(x+y+z)^5$? of $(x+y+z+w)^{10}$?
2. What is the coefficient of x^4 in the expansion of $(1+x+2x^{25})$?
3. Use Table 2.2 to write out the first ten terms in the expansions of
   (a) $(1+x)^{-2}$
   (b) $(1+2x)^{-4}$
   (c) $(2+3x)^{-3}$
4. Find the value of the sum $\sum \frac{8!}{i!j!k!}$, if i, j, k vary over all possible ordered triples of positive integers summing to 8.
5. Show that for n a positive integer $(-1)^n\binom{-n}{k-1} = (-1)^k\binom{-k}{n-1}$.
6. What is the number of shortest paths along grid lines from (0, 0, 0) to (2, 3, 4) in 3-space?
7. Use the binomial theorem to find $\sum \binom{n}{r} k^r$, k an arbitrary number.
8. Use the general form of the binomial theorem, $(x+y)^n = \sum \binom{n}{r} x^{n-r} y^r$, to prove the identity $2^n = \sum (-1)^{r+n} \binom{n}{r} 3^r$.

2.4. BINOMIAL PROBABILITIES

The binomial theorem comes up in solution of certain probability problems, such as the following:

> **Example 9.** Mary is to babysit for a family with four children. What is the probability that the children will all be boys? That they will be two boys and two girls?
>
> If we assume that boys and girls occur with probabilities .49 and .51, respectively, and that the genders of the four children are independent, we can reason as follows. The probability that a given child is a boy is .49, so the probability of all four being boys is $(.49)^4 = .057648$. The case of two boys and two girls is more complicated, because it can occur in more than one way. We might have BBGG or BGBG or ...? Clearly, the number of ways to have two boys and two girls is $\binom{4}{2} = 6$. Each of these ways, such as BBGG, has a probability of $(.49)^2(.51)^2 = .06245$, so the probability that some one of them will happen is $6(.06245) = .3747$.

This type of problem comes up so often that it has a special name. An experiment consists of *Bernoulli trials* (after James Bernoulli, 1654–1705) if:

(i) For each trial there are exactly two possible outcomes, *success* (S) and *failure* (F).
(ii) The probability of success is the same for each trial.
(iii) The trials are independent.

Example 9 fits this pattern if we consider each child in a family as a trial, with girl (say) being success, and boy failure. The method we used to find probabilities can be generalized as follows. Because of property **(i)**, the sample space of any such experiment is a cartesian product $\{S, F\} \times \{S, F\} \times \ldots$, with dimensional n equal to the number of trials in the experiment. Thus, (S, F, F, S) (equivalent to GBBG) is an outcome in our example. Clearly there are 2^n such outcomes in the sample space. If p is the probability of success and $q(= 1 - p)$ the probability of failure, then properties **(ii)** and **(iii)** assure that the probability of an outcome with k successes and $n - k$ failures is $p^k q^{n-k}$. But there are exactly $\binom{n}{k}$ ways in which such an outcome can occur in the sample space. Thus we have the following theorem.

Theorem 2.4. In an experiment consisting of n Bernoulli trials, with probability of success p, the probability of having exactly k successes is

$$\binom{n}{k} p^k q^{n-k}, \tag{2.19}$$

where $q = 1 - p$.

Example 10. An insurance company sells life insurance policies to six men, all the same age. Actuarial tables show that the probability that a man that age will be alive 20 years later is .60. What is the probability that in 20 years, exactly four of the men will be alive? That at least four will be alive?

Here we have $n = 6$, $p = .60$, $q = 1 - p = .40$, and $k = 4$, for the first question. Then the theorem gives

$$\binom{6}{4}(.60)^4(.40)^2 = 15 \times .1296 \times .16 = .31104$$

as the probability that exactly four will be alive. To answer the second question, we merely repeat this computation with $k = 5$ and $k = 6$, and add the results together. Thus the probability that at least four of the men will be alive is

$$\begin{aligned}\binom{6}{4}(.60)^4(.40)^2 + \binom{6}{5}(.60)^5(.40)^1 + \binom{6}{6}(.60)^6(.40)^0 \\ = .31104 + .186624 + .046656 \\ = .54432,\end{aligned}$$

or slightly more than half.

Example 11. Smallville's population is 58% Republican and 42% Democratic. A television crew selects five citizens at random for interviews. What is the probability that exactly one person selected is a Democrat? What is the probability that at least one Democrat is selected?

The answer to the first question is clearly $\binom{5}{1}(.42)^1(.58)^4$, which is about .2376464. For the second question, we could find the sum

$$\binom{5}{1}(.42)^1(.58)^4 + \binom{5}{2}(.42)^2(.58)^3 + \binom{5}{3}(.42)^3(.58)^2 + \binom{5}{4}(.42)^4(.58)^1 + \binom{5}{5}(.42)^5(.58)^0$$

but a moment's reflection makes us realize that the only way *not* to get even one Democrat is to get 5 Republicans, which happens with probability $\binom{5}{5}(.42)^0(.58)^5 = .0656357$. So the answer we seek is $1 - .0656357 = .9343643$.

The probabilities $\binom{n}{k}p^kq^{n-k}$ that we have been using are terms in the expansion of $(p + q)^n$, using the corollary to the binomial theorem,

$$(p + q)^n = \sum_{k=0}^{n} \binom{n}{k} p^k q^{n-k}. \tag{2.20}$$

This is certainly resonable, since the right-hand side merely adds the probabilities of all possible events (from no successes to n successes) while the left-hand side is just 1^n or 1. Thus the probabilities of all possible events add to 1, as they should.

In a series of Bernoulli trials the expectation often has more intuitive appeal than the probabilities we have been looking at. Recall that the expectation in any experiment is

$$e = p(E_1)P(E_1) + p(E_2)P(E_2) + \cdots + p(E_n)P(E_n),$$

where the E_i are all possible events and $p(E_i)$ and $\mathrm{P(E_i)}$ are the probability and the payoff, respectively, of event E_i. If we define the payoff of an event as the number of successful outcomes in it, we see that the expectation of our standard series of Bernoulli trials is

$$\sum_{k=0}^{n} k \binom{n}{k} p^k q^{n-k}. \tag{2.21}$$

This can be rewritten as follows:

$$\begin{aligned}
\sum_{k=0}^{n} k \frac{n!}{k!(n-k)!} p^k q^{n-k} &= \sum_{k=1}^{n} pn \frac{(n-1)!}{(k-1)!((n-1)-(k-1))!} p^{k-1} q^{n-k} \\
&= pn \sum_{k=1}^{n} \binom{n-1}{k-1} p^{k-1} q^{(n-1)-(k-1)} \\
&= pn \sum_{i=0}^{m} \binom{m}{i} p^i q^{m-i} \qquad (m = n - 1, i = k - 1)
\end{aligned}$$

The sum in this expression is equal to $(p + q)^m$, as in (2.20), which equals 1. Thus, the expectation is pn. Certainly it is reasonable that the expected number of successes is the probability of success (p) times the number of trials (n).

Example 12. George wants to get a World Series ticket, and he hears that 25% of all mail applications will be selected to receive a ticket. He figures that by sending in four applications, he will be sure to get a ticket. What is his actual probability of getting a ticket? What is the expected number of tickets he will receive? How many applications would he have to send in to raise his probability of success to at least 80%?

Clearly this is a series of Bernoulli trials, so the probability he will get k tickets is $\binom{4}{k}(.25)^k(.75)^{4-k}$. We could sum this for $k = 1, 2, 3$, and 4 to answer the first question, but it is easier to find the probability of getting *no* tickets, which is $\binom{4}{0}(.25)^0(.75)^4 = .3164$. Thus, his probability of getting at least one ticket is $1 - .31640625 = .6836$, not 1, as he thought.

To find the number of tickets he can expect to receive, we can calculate the probabilities of getting 0, 1, 2, 3, and 4 tickets, multiply and add, finding $.3164 \cdot 0 + .421875 \cdot 1 + .2109375 \cdot 2 + .046875 \cdot 3 + .00390625 \cdot 4 = 1$. The answer jolts us back to reality, making us remember that the expectation can be calculated much more readily as just $p \cdot n = .25 \cdot 4 = 1$. Thus, he can expect to receive one ticket, but he may receive more (up to four, in fact) or less. To raise his probability of success to at least 80%, he must lower his probability of failure (no ticket) below 20%. If $f(n)$ denotes the probability of failure, if he makes n applications, we found $f(4) = .3164$. Similarly, $f(5) = .2373$ and $f(6) = .178$. Thus, if he makes 6 applications, his probability of getting one or more tickets is $1 - .178 = .822$.

EXERCISES

1. Suppose a fair coin is flipped nine times. What is the probability it will be heads exactly five times? At least five times?

2. Suppose a fair die is thrown four times. What is the probability that a 6 will occur exactly twice? At most twice?

3. Two percent of the telephones coming off an assembly line are defective. If an inspector checks ten telephones, what is the probability that no defects will be noted?

4. If a machine making metal stampings produces a defective stamping with probability .2, is it more likely that an inspector will find no defects among ten sampled stampings, or at most one defect among twenty sampled stampings?

5. In a ten-question true–false exam, a student must achieve six correct answers to pass. If he selects his answers randomly, what is the probability he will pass?

6. A bridge hand is 13 cards from a 52-card deck. What is the probability that a bridge player will receive no aces in five successive hands?

7. Helen is to throw a fair coin N times, and receive a prize if she throws exactly four heads. If she can choose N, what number should she choose, and what is the probability she will win with that choice of N?

8. What is the probability that the nth head will occur after exactly n tails have been thrown, in tossing a fair coin?

9. What is the probability that in three throws of a fair die:
   (a) No 2 appears?
   (b) One 2 appears?
   (c) Two 2's appear?
   (d) Three 2's appear?
   (e) What is the sum of (a) to (d)?

10. If an unfair coin comes up heads with probability .7, what is the probability of getting exactly five heads in nine flips? At least five heads?

11. An urn contains six red balls and ten white balls. You reach in and pull out five balls. For each red ball you get, you receive $1.00. What is your expected payoff from this game? (Leave answer in expanded form.)

2.5. REMARKS

We have seen that the binomial coefficients, introduced in Chapter 1, appear in the binomial theorem, Pascal's Triangle, many identities, and probabilities. They are fundamental also in some of the further counting we do in Chapter 3.

Although the binomial theorem appears in various forms in early manuscripts of several mathematical disciplines, it was first carefully stated and proved by James Bernoulli in the late seventeenth century. In fact, his book, *Ars Conjectandi* (1713) gathered many basic combinatorial methods together for the first time. The binomial theorem for negative and fractional exponents was worked out by Newton when he was an undergraduate at Cambridge. Two of his letters describing this work are presented in a modern collection [1]. They provide an unusual glimpse of the notation and thinking of one of the inventors of calculus.

The numerical properties of binomial and multinomial coefficients have been studied extensively in number theory (see Dickson [2], Chapter 9). An entertaining treatment of binomial probabilities can be found in Packel [3]. The introductory texts mentioned at the end of Chapter 1 contain various approaches to the binomial theorem and its consequences.

[1] J. R. Newman, *The World of Mathematics*, Vol. 1, Simon and Schuster, New York, 1956, pp. 521–524.

[2] L. E. Dickson, *History of the Theory of Numbers*, Vol. 1, Chelsea, New York, 1952.

[3] E. Packel, *The Mathematics of Games and Gambling*, MAA, New York, 1981.

CHAPTER 3

Enumeration

We have already examined in some detail certain special counting problems, especially those having to do with the number of ways of choosing objects from sets. In this chapter we will examine some other counting methods, and give special emphasis to a phenomenon that is very common in combinatorics—the underlying similarities between seemingly distinct problems.

3.1. FIBONACCI NUMBERS

In 1202 the Italian mathematician Fibonacci posed the following problem:

> Each month, beginning when she is two months old, the female of a pair of rabbits gives birth to a pair of rabbits (of different sexes). Beginning two months later, the female of the new pair will also begin producing a new pair each month, and so on.... Find the number of rabbits at the end of the year if there is one new pair of rabbits at the beginning of the year.

Fibonacci was the greatest European mathematician before the Renaissance, and intended the rabbit problem as an exercise in arithmetic rather than as a practical biological example. The answer can be obtained by direct figuring, which is simplified by the following observations. Clearly each month all rabbits alive the previous month are still alive (we assume no dead rabbits). Also, there is one new pair of rabbits for each pair of rabbits that were alive *two* months before. Thus, if $F(n)$ denotes the number of pairs after n months have passed, we have $F(n) = F(n-1) + F(n-2)$. The numbers $F(n)$ are known as the Fibonacci numbers, and occur in many settings, as we shall see. The equation we give for $F(n)$ is called a linear recurrence relation of order 2. Note that the formula does not determine $F(n)$ uniquely unless two values of F are given. In this case we have $F(0) = 1$ (one initial pair) and $F(1) = 1$ (still no offspring). With that information, we can easily find that the first few values of F are 1, 1, 2, 3, 5, 8, 13, 21, 34, 55, 89, 144, ... But what is $F(500)$? We will postpone a general discussion of recursion relations and methods of actually solving for $F(n)$ until later in the chapter.

Consider now the following example.

Example 1. In how many ways can eight plus signs and three minus signs be lined up in a row so that no two minus signs are adjacent?

This example is easy to solve with what we know from Chapter 1 if viewed as follows. Write down the eight plus signs together with nine x's as follows:

$$x + x + x + x + x + x + x + x + x.$$

Then, to place three minus signs as required, all we need to do is replace three of the x's by minus signs, and drop the others. There are nine x's so there are clearly $\binom{9}{3}$ ways to get such a sequence of + and − signs. With n plus signs and k minus signs, the number is clearly $\binom{n+1}{k}$. Hence we have the following theorem.

Theorem 3.1. The number of ways in which n plus signs and k minus signs can be lined up so that no two minus signs are adjacent is $\binom{n+1}{k}$.

A more difficult question is the following:

Example 2. Suppose we have ten x's in a row, and each may be replaced by a + or a −, but we must not place two minus signs together. How many ways can that be done?

Note that with no restrictions on placement each x can be replaced two ways, so there are 2^{10} ways to replace them all. With the restriction, however, we must break the problem up into cases, which are each like the problem we just solved. We may use ten plus signs, or nine plus signs and one minus sign, eight and two, and so on. Thus, by our previous result the number of ways to do this is

$$\binom{11}{0} + \binom{10}{1} + \binom{9}{2} + \binom{8}{3} + \binom{7}{4} + \binom{6}{5}.$$

The other terms [$\binom{5}{6}$, etc.] are zero, and the answer we get is 144.

It is interesting to notice that $F(11)$ is also 144. This is not a coincidence, as we now show. Let us denote the number of ways of making an n-sequence of $+$ and $-$, with no two minus signs together, by $S(n)$. A sequence of length 1 can be made in two ways, so $S(1) = 2$. Also, $S(2) = 3$. To find $S(n)$ in general, note that if an n-sequence begins with a plus sign, the remaining $n - 1$ sequence can be chosen arbitrarily, that is, in $S(n - 1)$ ways. However, an n-sequence beginning with a minus sign necessarily has a plus sign next, and may then continue arbitrarily in $S(n - 2)$ ways. Thus we see $S(n) = S(n - 1) + S(n - 2)$, and we have the Fibonacci recurrence. In fact, $S(n) = F(n + 1)$ because the sequences have different initial values. Note that $S(1) = 2 = F(2)$, $S(2) = 3 = F(3)$, $S(3) = S(2) + S(1) = F(3) + F(2) = 5$, and so on. Hence we have the next theorem.

Theorem 3.2. If $S(n)$ denotes the number of n-sequences of $+$ and $-$ with no two minus signs together then

$$S(n) = S(n - 1) + S(n - 2) = F(n + 1).$$

But we already discovered another way to calculate $S(n)$, namely,

$$S(n) = \sum_{k=0}^{c} \binom{n + 1 - k}{k},$$

where $c = [(n + 1)/2]$, and $[x]$ denotes the largest integer not exceeding x. Thus we have a formula for $F(n)$.

Corollary 3.1. $F(n) = \sum_{k=0}^{d} \binom{n-k}{k}$, where $d = [\frac{n}{2}]$

EXERCISES

1. What is the answer to Fibonacci's rabbit problem?
2. Find all n such that $F(n) = n^2$.
3. Show $F(n) = 2F(n - 2) + F(n - 3) = 3F(n - 3) + 2F(n - 4)$. Can you generalize this?
4. Show $F(n + m + 1) = F(m)F(n + 1) + F(m - 1)F(n)$.
5. Use exercise 4 to prove that $F(nk + 1)$ is a multiple of $F(k)$.
6. Prove inductively that $\begin{pmatrix} 1 & 1 \\ 1 & 0 \end{pmatrix}^n = \begin{pmatrix} F(n+1) & F(n) \\ F(n) & F(n-1) \end{pmatrix}$.

7. Use exercise 6 to prove that $F(n+1)F(n-1) - F^2(n) = (-1)^n$.

8. Show that, with one exception, $F(n)$ is prime only if n is prime.

9. Is $F^2(n) + F^2(n+1)$ always a Fibonacci number?

10. Show $F(0) + F(1) + \cdots + F(n) = F(n+2) - 1$.

11. Show $F(0) + F(2) + F(4) + \cdots + F(2n) = F(2n+1)$.

12. In how many ways can ten plus signs and nine minus signs be placed in a row satisfying the following conditions?
    (a) No restriction.
    (b) No two plus signs are adjacent.
    (c) No two minus signs are adjacent and no two plus signs are adjacent.

13. How many 8-place binary strings have no 00 substrings?

14. You are to fill a fixed box of size $2 \times n$ with 1×2 rectangles. For a given n this can be done in $B(n)$ ways. [For example, $B(1) = 1$, $B(2) = 2$, $B(3) = 3$.] Find a recursion for the numbers $B(n)$.

15. A bricklayer is laying a row of twenty bricks.
    (a) If he has a large supply of red bricks and a large supply of white bricks, how many ways can he do the job?
    (b) If he has thirteen red bricks and seven white bricks, how many ways can he do the job?
    (c) If he has fourteen red bricks and eight white bricks, how many ways can he do the job without laying two white bricks side by side?

3.2. LINEAR DIOPHANTINE EQUATIONS

We begin with a simple example.

Example 3. How many solutions are there in positive (nonzero) integers for the equation $x + y + z = 5$?

The number of solutions is so small that they may be listed easily. In fact the possibilities for (x, y, z) are $(3, 1, 1)$, $(2, 2, 1)$, $(2, 1, 2)$, $(1, 3, 1)$, $(1, 2, 2)$, and $(1, 1, 3)$, so there are just six solutions. We consider the first and last solutions distinct because a different variable takes on the value 3. The problem of counting solutions that are distinct in an unordered sense, so that, for example, the first and last solutions are the same, will be considered later, when we discuss partitions of an integer.

There is an easy method for finding the number of positive solutions to such equations, suggested by the following equation:

$$1 + 1 + 1 + 1 + 1 = 5.$$

To select a solution (x, y, z) we need only choose two of the plus signs to break the sum up into three parts. [Thus $1 \oplus 1 + 1 + 1 \oplus 1 = 5$ corresponds to $1 + 3 + 1 = 5$, or the solution (1,3,1).] This can be done in $C(4, 2) = 6$ ways. Generalizing this example we have the following theorem.

Theorem 3.3. The equation

$$x_1 + x_2 + x_3 + \cdots + x_n = m$$

has exactly $\binom{m-1}{n-1}$ solutions in positive integers.

Proof. Each choice of $n - 1$ plus signs from the $m - 1$ plus signs in a sum of m 1's corresponds to a solution in positive integers.

Example 4. How many solutions are there, in positive integers, for the equation $x_1 + x_2 + x_3 + x_4 = 16$?

Here $m = 16$, $n = 4$, so the answer is $\binom{15}{3} = 455$.

To find the number of solutions in non-negative integers to our equation of Example 3, a different analysis is required. We again write our five 1's, but now we leave spaces, and insert two plus signs anywhere. Any arrangement of five 1's and two plus signs will suggest a solution, in non-negative integers, to the original equation. For example, 1 1 + + 1 1 1 suggests $2 + 0 + 3$, or (2, 0, 3) as a solution. Again 1 + 1 1 1 + 1 suggests a solution (1, 3, 1). The number of solutions suggested in this way is the number of permutations of seven things, five of one kind and two of the other, which is $7!/5!2! = 21$. Note this is also $\binom{7}{2}$. Can you see why? In general, we have the following theorem.

Theorem 3.4. The equation

$$x_1 + x_2 + \cdots + x_n = m$$

has exactly $\binom{m+n-1}{n-1}$ solutions in non-negative integers.

Proof. This binomial coefficient counts the number of ways of placing

m 1's and $n - 1$ plus signs in a row, and such placements correspond to solutions in non-negative integers as outlined above.

Example 5. How many solutions are there in non-negative integers for the equation $x_1 + x_2 + x_3 + x_4 = 16$?

Here $m = 16$, $n = 4$, so the theorem gives $\binom{16+4-1}{4-1} = \binom{19}{3} = 969$.

Finally, suppose we are interested in restricted solutions to an equation. For example, suppose we require the number of solutions in integers x, y, and z, each greater than 2, to the equation $x + y + z = 16$. This problem can be reduced to a previous one by setting $q = x - 2$, $r = y - 2$, and $s = z - 2$. Then the original equation becomes $(q + 2) + (r + 2) + (s + 2) = 16$, or $q + r + s = 10$. But x, y, and z are each greater than 2 if and only if q, r, and s are each positive, so we need only find the number of solutions to our new equation in positive integers q, r, and s. Hence the answer is $\binom{9}{2} = 36$. Clearly various restrictions on the variables can be handled in this way, as shown in the following exercises. In fact, the answer we derived for solutions in non-negative integers can be derived in this way from the answer for the case of positive integers.

Example 6. How many solutions are there to the equation $x + y + z + w = 19$, in which x, y, z, and w are integers, x is positive, y is at least 7, and z and w are non-negative?

Let $q = x$, $r = y - 6$, $s = z + 1$, $t = w + 1$. Then the equation $q + (r + 6) + (s - 1) + (t - 1) = 19$ is equivalent to the original equation, and the conditions on x, y, z, and w translate to a requirement that q, r, s, and t are positive. This simplifies to $q + r + s + t = 15$, which has $\binom{14}{3} = 364$ positive solutions.

To handle restrictions of the form $a \leqslant x_i \leqslant b$, we need to develop more theory, which we proceed to do in the next section.

EXERCISES

Find the number of solutions to the following equations in integers satisfying the given restraints.

1. $x + y + z = 7$ $\qquad x \geqslant 0, y \geqslant 0, z \geqslant 0$

2. $x + y + z + w = 14$ $\qquad x > 0, y > 0, z > 0, w > 0$

3. $x + y + z = 9$ $\qquad x > 0, y > 2, z > 1$

4. $x + y + z = 10$ $\quad x > -1, y > -2, z > -3$

5. $x + y - z = 9$ $\quad x > 0, y > 0, z < 0$

6. $x + y - z = 7$ $\quad x > 0, y > 2, z < 2$

7. $2x - 2y + 2z = 10$ $\quad x \geqslant 1, y \leqslant 1, z \geqslant 0$

8. $2y + z = 17$ $\quad y \geqslant 1, z \geqslant 1$

3.3. INCLUSION – EXCLUSION

We introduce this important counting method with an example.

Example 7. In a certain sports club with 54 members, 34 members play tennis, 22 play golf, and 10 play both sports. How many members play neither tennis nor golf?

To solve the problem, we can partition the members into classes without common elements. One class is those who play both sports, one is those who play just tennis, one is those who play just golf, and one is those who play neither sport. We are told the first class contains 10 people. From that and the fact that 34 play tennis, we see that 24 play tennis alone. Similarly, 12 play only golf. Our information so far indicates that $10 + 24 + 12 = 46$ play one or both of the sports, so 8 play neither one, solving our original problem.

This solution can be illustrated as in Figure 3.1(a). A more complicated version of the same problem is as follows.

Example 8. Suppose we consider the same sports club, but we are now told that there are 11 members who play handball, of whom 6 play handball and tennis, 4 play handball and golf, and 2 play all three sports. How many people play none of the three sports?

The simplest way to analyze the problem is with a diagram, as shown in Figure 3.1(b), which gives the number 5 for the people who play none of the sports. This can be found as follows. Let A, B, and C be the sets of people who play tennis, golf, and handball, respectively. Then to count the people *not* in A, B, or C, we first remove from the set S of all people, those people in A, B, and C. Thus we have

$$|S| - |A| - |B| - |C|.$$

This is not the right number because, those people who are in

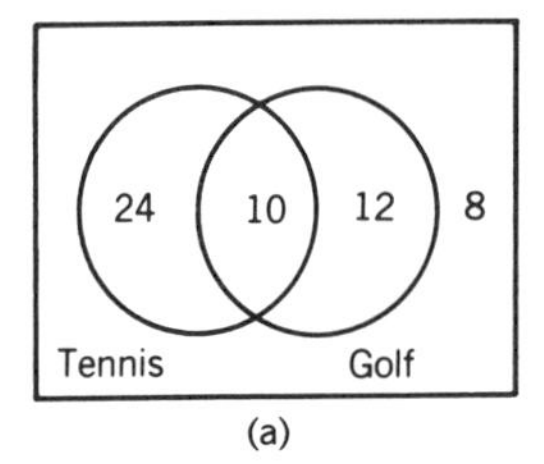

(a)

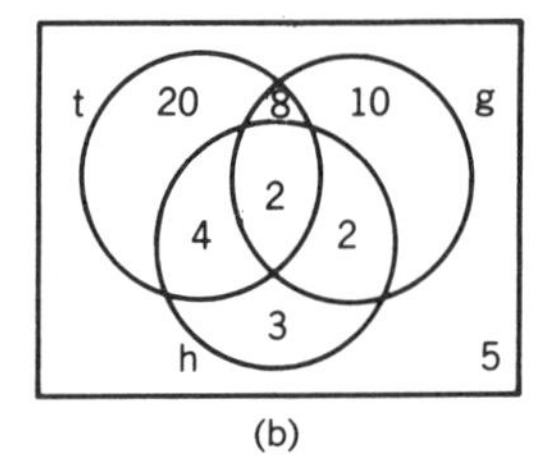

(b)

Figure 3.1

set A as well as set B (i.e., in $A \cap B$) have been subtracted twice. Thus we should use

$$|S| - |A| - |B| - |C| + |A \cap B| + |A \cap C| + |B \cap C|,$$

where the last three terms add back in (once) those people who were removed twice. But this is still not correct, as we see by considering a person who lies in A, B, and C (i.e., in $A \cap B \cap C$). He or she is counted in S, subtracted three times, and then added back three times. Thus, this expression counts him or her, whereas we only want to count those people who are in S but *not* in A, B, or C. Thus we use

$$|S| - |A| - |B| - |C| + |A \cap B| + |A \cap C| + |B \cap C| - |A \cap B \cap C|.$$

In the example of Figure 3.1(b) this is

$$54 - 34 - 22 - 11 + 10 + 6 + 4 - 2 = 5.$$

This principle can be formulated in general as follows. Suppose we have a set of N objects and a set of n properties $\alpha_1, \alpha_2, \ldots, \alpha_n$. Some of the N objects may have none of the properties, some may have one or more of them. We use the symbol $N(\alpha_i \alpha_j \ldots \alpha_k)$ to denote the number of objects having properties $\alpha_i, \alpha_j, \ldots, \alpha_k$ (and perhaps other properties as well). To emphasize when we are counting objects that lack a particular property, we use a prime on the appropriate α. Thus $N(\alpha_1 \alpha'_2)$ is the number of objects that have property α_1 (and perhaps other properties) but *not* property α_2. Thus $N(\alpha'_1 \alpha'_2 \ldots \alpha'_n)$ denotes the number of objects with none of the properties α_i.

The principle of inclusion and exclusion (also known as the sieve principle) states the following,

Theorem 3.5.

$$\begin{aligned} N(\alpha_1'\alpha_2'\cdots\alpha_{n-1}'\alpha_n') = {} & N - N(\alpha_1) - N(\alpha_2) - \cdots - N(\alpha_n) \\ & + N(\alpha_1\alpha_2) + N(\alpha_1\alpha_3) + \cdots + N(\alpha_{n-1}\alpha_n) \\ & - N(\alpha_1\alpha_2\alpha_3) - N(\alpha_1\alpha_2\alpha_4) - \cdots - N(\alpha_{n-2}\alpha_{n-1}\alpha_n) \\ & + \quad \cdots \qquad \cdots \qquad \cdots \\ & + (-1)^n N(\alpha_1\alpha_2\cdots\alpha_{n-1}\alpha_n). \end{aligned} \tag{3.1}$$

Proof. The proof is by induction on the number of properties, that is, on the number n. If $n = 1$ the formula is $N(\alpha_1') = N - N(\alpha_1)$, which is obviously true. Suppose (3.1) is true for $n - 1$ properties. Using the formula on the objects that have a property α_n, we obtain.

$$\begin{aligned} N(\alpha_1'\alpha_2'\ldots\alpha_{n-1}'\alpha_n) = {} & N(\alpha_n) - N(\alpha_1\alpha_n) - N(\alpha_2\alpha_n) - \cdots - N(\alpha_{n-1}\alpha_n) \\ & + N(\alpha_1\alpha_2\alpha_n) + N(\alpha_1\alpha_3\alpha_n) + \cdots + N(\alpha_{n-2}\alpha_{n-1}\alpha_n) \\ & + \qquad \cdots \qquad \cdots \qquad \cdots \qquad \cdots \\ & + (-1)^{n-2}(\alpha_1\alpha_2\ldots\alpha_{n-2}\alpha_n) + \cdots \\ & + (-1)^{n-2}(\alpha_2\alpha_3\ldots\alpha_{n-1}\alpha_n) \\ & + (-1)^{n-1}(\alpha_1\alpha_2\ldots\alpha_{n-1}\alpha_n). \end{aligned} \tag{3.2}$$

We also have

$$\begin{aligned} N(\alpha_1'\alpha_2'\ldots\alpha_{n-1}') = {} & N - N(\alpha_1) - N(\alpha_2) - \cdots - N(\alpha_{n-1}) \\ & + N(\alpha_1\alpha_2) + N(\alpha_1\alpha_3) + \cdots + N(\alpha_{n-2}\alpha_{n-1}) \\ & + \cdots \\ & + (-1)^{n-1}(\alpha_1\alpha_2\alpha_3\ldots\alpha_{n-1}). \end{aligned} \tag{3.3}$$

Subtracting (3.2) from (3.3), we get the formula (3.1) if we note that

$$N(\alpha_1'\alpha_2',\ldots,\alpha_{n-1}') - N(\alpha_1',\alpha_2',\ldots,\alpha_{n-1}',\alpha_n) = N(\alpha_1'\alpha_2'\ldots\alpha_{n-1}'\alpha_n').$$

Thus assuming the principle of inclusion and exclusion for $n - 1$, we have derived it for n.

Example 9. How many solutions are there to the equation $x + y + z + w = 20$ in positive integers $x \leqslant 6$, $y \leqslant 7$, $z \leqslant 8$, $w \leqslant 9$?

To use inclusion–exclusion, we let the objects be solutions (in positive integers) to this equation. Say a solution has property α_1 if $x > 6$, property α_2 if $y > 7$, property α_3 if $z > 8$, and property α_4 if $w > 9$. Then the answer to the question asked in the problem is the number $N(\alpha_1'\alpha_2'\alpha_3'\alpha_4')$

The total number of solutions (in positive integers) to the equation given is $\binom{19}{3}$, by the theory of the previous section. Thus $N = \binom{19}{3}$. To find $N(\alpha_1)$ we again use the theory of that section, and have $N(\alpha_1) = \binom{20-6-1}{4-1} = \binom{13}{3}$. Continuing we find $N(\alpha_2) = \binom{12}{3}$, $N(\alpha_3) = \binom{11}{3}$, $N(\alpha_4) = \binom{10}{3}$, $N(\alpha_1\alpha_2) = \binom{6}{3}$, $N(\alpha_1\alpha_3) = \binom{5}{3}$, and so on. By inclusion and exclusion we obtain

$$\begin{aligned} N(\alpha_1'\alpha_2'\alpha_3'\alpha_4') &= \binom{19}{3} - \binom{13}{3} - \binom{12}{3} - \binom{11}{3} - \binom{10}{3} \\ &\quad + \binom{6}{3} + \binom{5}{3} + \binom{4}{3} + \binom{4}{3} + \binom{3}{3} \\ &= 969 - 286 - 220 - 165 - 120 \\ &\quad + 20 + 10 + 4 + 4 + 1 \\ &= 217. \end{aligned}$$

We note that many of the terms $N(\alpha_1'\alpha_2'\ldots)$ are zero.

Example 10. A permutation $a_1a_2 \ldots a_n$ of the numbers 1 to n in which no number occurs in its natural position (i.e., $a_i \neq i$ for all i) is a *derangement*. Thus 231 is a derangement, while 321 is not, because 2 is in its natural position. Find D_n, the number of derangements of the numbers 1 to n.

Here we let the set of all permutations of 1 to n be our objects, so $N = n!$. Property α_i is that the number i occurs in the ith position of the permutation. Then $N(\alpha_1'\alpha_2'\ldots\alpha_n')$ is the number of derangements. Note that $N(\alpha_i)$ is the number of permutations of 1 to n with i in the ith position, which is clearly $(n-1)!$. $N(\alpha_i\alpha_j)$ counts permutations with i and j in the ith and jth spots, respectively, leaving us free to place the other $n-2$ numbers, so $N(\alpha_i\alpha_j) = (n-2)!$. Continuing in this way, clearly $N(k$ properties$) = (n-k)!$. Thus, by inclusion and exclusion we find

$$\begin{aligned} N(\alpha_1'\alpha_2'\ldots\alpha_n') &= n! - n(n-1)! + \binom{n}{2}(n-2)! - \binom{n}{3}(n-3)! \\ &\quad + \cdots + (-1)^n\binom{n}{n}0! \\ &= n!\left(1 - \frac{1}{1!} + \frac{1}{2!} - \frac{1}{3!} + \cdots + (-1)^n\frac{1}{n!}\right). \end{aligned}$$

The expression in parentheses should look familiar because it is the beginning of the Taylor series expansion for e^{-x}, evaluated at $x = 1$. It is very close to e^{-1}, or 0.36788, even for small n.

The exercises suggest many other ways to use this principle.

EXERCISES

1. Find the number of integral solutions to the equation $x + y = 10$, obeying the restraints $0 \leqslant x \leqslant 6, 0 \leqslant y \leqslant 10$.

2. Find the number of integral solutions to the equation $x + y + z = 13$, obeying the restraints $1 \leqslant x \leqslant 8, -2 \leqslant y \leqslant 2, 1 \leqslant z$.

3. A bookbinder is to bind twelve different books in red, green, and brown cloth. In how many ways can he do this, if each color of cloth is to be used for at least one book?

4. In how many ways can the letters of the word zigzag be permuted so that the same letters do not appear together?

5. How many non-negative integers smaller than 1,000,000 include all four of the digits 1, 2, 3, 4?

6. How many numbers from 0 to 999 are not divisible by either 5 or 7?

7. How many 3-digit sequences have no digit twice in a row? Same question for n-digit sequences.

8. Eight people enter an elevator. At each of four floor stops at least one person leaves the elevator. After four floor stops the elevator is empty. In how many ways can this be done?

9. Three Egyptians, three Israelis, and one American are to be seated in a row so that no three countrymen sit together. In how many ways can that be done? What if no two countrymen sit together?

10. If ten people check their hats, and the drunken hat-check girl returns the hats randomly to the people, what is the probability that no one gets the right hat?

11. Each of twelve students is taking an examination on two subjects. One professor examines the students in one subject and another in the other subject, and each professor takes five minutes of the hour from 1 to 2 p.m. to examine each student. In how many ways can the examinations be scheduled so that no student has both exams at once?

12. How many six-digit numbers contain exactly three different digits?

$3+2+1+1=7$ $\quad$ $4+2+1=7$

Figure 3.2

3.4. PARTITIONS

In considering positive integral solutions to equations such as $x + y + z = 5$, we agreed to call the solutions (3, 1, 1) and (1, 1, 3) distinct. If we decide instead to concern ourselves only with the summands, disregarding their order, we have a partition problem. To be precise, an *unordered partition* of n is a collection of positive integers that sum to n. The partitions of 5 are 5, $4 + 1$, $3 + 2$, $3 + 1 + 1$, $2 + 2 + 1$, $2 + 1 + 1 + 1$, and $1 + 1 + 1 + 1 + 1$. Thus $P(5)$, the number of partitions of 5, is 7. Each summand is a *part* of n. It is very difficult to find a formula for $P(n)$. We shall discover how to find $P(n)$ in the next section, but for now we merely point out some elementary properties of partitions. A most useful tool in such investigations is *Ferrer's diagram*, which represents a partition geometrically. The examples shown in Figure 3.2 should be sufficient to explain the concept. The two partitions in Figure 3.2 are said to be *conjugate* because the diagram of one changes to that of the other when rows are interchanged with columns. This geometric correspondence yields several theorems, the simplest of which is the following.

Theorem 3.6. The number of partitions of n into m parts is equal to the number of partitions of n into parts, the largest of which is m.

A slightly more sophisticated application is the following.

Theorem 3.7. The number of partitions of n into parts that are odd and unequal is the same as the number of partitions of n whose Ferrer's diagrams are conjugate with themselves.

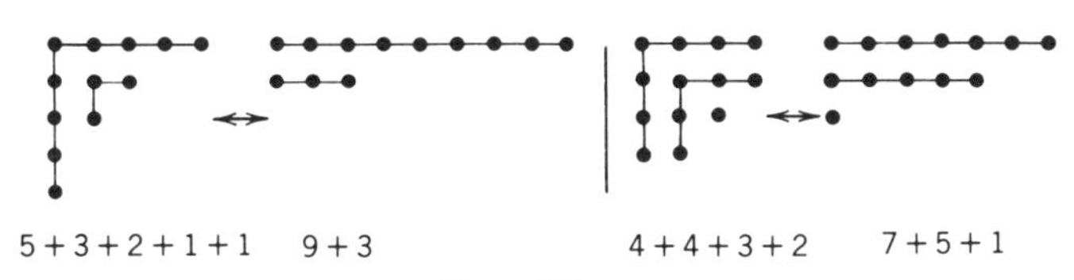

Figure 3.3

Proof. We simply show a 1–1 correspondence between these two kinds of partitions, via their Ferrer's diagrams (see Figure 3.3).

Finally, we mention perfect partitions. A partition of n is *perfect* if every integer $k \leqslant n$ is representable in exactly one way from the parts of n. Here identical parts are considered indistinguishable so that, for example, $1 + 1 + 1 + \cdots + 1$ is a perfect partition. The perfect partitions of 7 are $4 + 1 + 1 + 1$, $4 + 2 + 1$, $2 + 2 + 2 + 1$, and $1 + 1 + 1 + 1 + 1 + 1 + 1$. Note that $5 + 1 + 1$ is not perfect since 3 cannot be represented as a sum, and $2 + 2 + 1 + 1 + 1$ is not perfect since 3 can be represented as $2 + 1$ or as $1 + 1 + 1$. We cannot find the perfect partitions of n but we can derive a rather surprising fact about their number.

Theorem 3.8. The number of perfect partitions of n is the same as the number of ordered factorizations of $n + 1$ into numbers $\geqslant 2$.

Proof. Since $1 \leqslant n$, 1 must be a part of any perfect partition. Say 1 occurs $m_1 - 1$ times. Then m_1 must be the next part. Say m_1 occurs $m_2 - 1$ times. Then after the 1's and the m_1's, all numbers less than $m_1 m_2$ are expressible. Thus the next part must be $m_1 m_2$, and it occurs $m_3 - 1$ times (say). Continuing in this way, we partition n and find $n = (m_1 - 1) + (m_2 - 1)m_1 + (m_3 - 1)m_1 m_2 + \cdots + (m_k - 1)(m_1 m_2 \ldots m_{k-1}) = m_1 m_2 \ldots m_k - 1$. Thus our perfect partition has yielded an ordered factorization of $n + 1$, and $m_i \geqslant 2$ for all i. Since the ordered factorization $m_1 m_2 \ldots m_k$ determines, in turn, a perfect partition of n, we are done.

Example 11. Write down all perfect partitions of 11 (see Figure 3.4).

Ordered Factorizations of 12	Perfect Partitions of 11
$6 \cdot 2$	$1 + 1 + 1 + 1 + 1 + 6$
$4 \cdot 3$	$1 + 1 + 1 + 4 + 4$
$3 \cdot 4$	$1 + 1 + 3 + 3 + 3$
$2 \cdot 6$	$1 + 2 + 2 + 2 + 2 + 2$
$3 \cdot 2 \cdot 2$	$1 + 1 + 3 + 6$
$2 \cdot 3 \cdot 2$	$1 + 2 + 2 + 6$
$2 \cdot 2 \cdot 3$	$1 + 2 + 4 + 4$

Figure 3.4

EXERCISES

1. Find the conjugates of the following partitions:
   (a) $7 = 4 + 1 + 1 + 1$
   (b) $9 = 6 + 2 + 1$

(c) $15 = 5 + 5 + 4 + 1$
(d) $13 = 4 + 3 + 2 + 2 + 1 + 1$

2. For each of the self-conjugate partitions in exercise 1, find the corresponding odd, unequal partition.

3. For the partition $9 + 7 + 3 + 1$ find the corresponding self-conjugate partition.

4. Suppose partitions of n are always written as sums of nonincreasing terms $n = n_1 + n_2 + n_3 + \cdots + n_k$. Show that the number of partitions of n into unequal parts is equal to the number of partitions of n in which $|n_i - n_j| \leqslant |i - j|$, for all i, j, and the smallest part is 1.

5. Compare the number of partitions of n into an odd number of parts with the number of partitions of n into odd parts.

6. Compare the number of partitions of n into odd parts with the number of partitions of n into even parts.

7. When is the trivial partition $1 + 1 + 1 + \cdots + 1$ the only perfect partition of n?

8. Find all perfect partitions of 8.

9. Show that every part of a perfect partition of n is a divisor of $n + 1$.

10. (a) Show that the number of solutions, in non-negative integers, of $x + 2y + 3z + 4w = 4$ is the number of partitions of 4.
(b) Show that the number of solutions, in non-negative integers, of $x_1 + 2x_2 + \cdots + nx_n = n$ is $P(n)$.

11. Show that the number of partitions of n into even parts is equal to the number of partitions of n in which every part appears an even number of times.

12. Show that the number of partitions of n that contain exactly three distinct parts is $[(n^2 - 6n + 12)/12]$. [*Hint*: use inclusion–exclusion.]

13. If n has prime factorization $n = p_1^{a_1} p_2^{a_2} \ldots p_k^{a_k}$, show that the number of perfect partitions of $n - 1$ is $\binom{a_1 + a_2 + \cdots + a_k}{a_1, a_2, \cdots, a_k}$.

3.5. GENERATING FUNCTIONS

The idea of generating functions is extremely general, but we will restrict ourselves to the generating functions known as ordinary ones, ignoring exponential and other esoteric types. If $\{a_n\}$ is any sequence of real numbers (usually positive integers), then $f(x)$ is the *generating function* for $\{a_n\}$ if

$$f(x) = a_0 + a_1 x + a_2 x^2 + \cdots + a_n x^n + \cdots.$$

For example, since

$$\frac{1}{1-x} = 1 + x + x^2 + x^3 + \cdots \tag{3.4}$$

we say that $f(x) = 1/(1-x)$ is the generating function for the sequence $\{1, 1, 1, \ldots\}$. Also, by the binomial theorem we know that

$$\frac{1}{1-2x} = 1 + 2x + 2^2 x^2 + \cdots + 2^n x^n + \cdots \tag{3.5}$$

so that $f(x) = 1/(1-2x)$ is the generating function for $\{1, 2, 2^2, \ldots, 2^n, \ldots\}$.

In considering sums of this type we are not concerned with questions of convergence, and are using x only as a formal mark, not as a variable for which values may be substituted. If the value 1 were substituted for x in the equations above, one would give trouble and the other would give nonsense. Other familiar generating functions are:

(a) $(1+x)^n$ generates $\{\binom{n}{0}, \binom{n}{1}, \ldots, \binom{n}{n}, \binom{n}{n+1}, \ldots\}$.
(b) 1 generates $\{1, 0, 0, \ldots\}$.
(c) $(1-x)^{-2}$ generates $\{1, 2, 3, \ldots, n, \ldots\}$.

Example (c) can be derived by the binomial theorem or more simply by taking the derivative of both sides of equation (3.4) above. In general, if f is the generating function for $\{a_n\}$ and f and all of its derivatives exist at $x = 0$, then a_n can be found from f since $n!a_n = f^{(n)}(0)$. This method of deriving a_n is seldom practical, however.

Generating functions are used for many purposes in combinatorics. In particular, enumeration problems often involve derivation of a sequence of numbers [the numbers $F(n)$ and $P(n)$ spring to mind]. The best solution is a formula for the desired numbers, but when this cannot be found either a recurrence or a generating function involving the sequence is an acceptable partial solution since each one determines the sequence in principle, if not in a practical way.

We shall derive generating functions for $P(n)$ and $F(n)$ in very different ways and then show how generating functions sometimes can be used to solve recursion relations.

Perhaps a trivial example best illustrates one way of arriving at generating functions. Suppose you are to choose a sundae and are allowed chocolate, vanilla, or strawberry ice cream, each with either fudge or marshmallow topping. Application of elementary counting techniques yields $3 \cdot 2 = 6$ as the number of different sundaes that can be chosen. Writing this as $(1 + 1 + 1)(1 + 1) = 1 + 1 + 1 + 1 + 1 + 1$ may seem silly, but it is instructive. In a product of factors of all possible pairs, one member from the first factor and one member from the second are listed. This is an exact parallel to what we want to do: Find all combinations of ice cream and topping. The product $(c + v + s)(f + m) = cf + cm + vf + vm + sf + sm$ is even more interesting. because it actually identifies each type of sundae rather than just giving us a number.

In this way of thinking a sum lists choices, and a product is used to indicate a combination of choices. Thus + means "or" and · means "and". We can apply this idea to get the generating function for the number $P(n)$ of partitions of an integer n. Recall that the generating function for $P(n)$ is a sum of powers of x in which the coefficient of x^n is $P(n)$. To form a partition we choose some number of 1's (none, one, two, etc.), then some number of 2's, and so on, in such a way that the numbers chosen sum to n. The following product indicates all of those choices:

$$(1 + x + x^2 + \cdots)(1 + x^2 + x^4 + \cdots)(1 + x^3 + x^6 + \cdots)\ldots. \tag{3.6}$$

Here our choice of one term from the first factor indicates the number of 1's we will have, in the second factor we choose 2's (e.g., x^6 indicates a choice of three 2's), and so on. We will get a contribution of 1 to the coefficient of x^n exactly when we choose terms whose exponents sum to n. But the terms chosen will then correspond to a particular partition of n. For example, $x^3 \cdot x^2 \cdot 1 \cdot x^8 = x^{13}$ corresponds to $1 + 1 + 1 + 2 + 4 + 4 = 13$. Thus the coefficient of x^n in (3.6), which can also be expressed as

$$\prod_{k=1}^{\infty} (1 - x^k)^{-1} = \frac{1}{(1 - x)(1 - x^2)(1 - x^3)\ldots}, \tag{3.6a}$$

is $P(n)$, if the convention $P(0) = 1$ is employed. This is an extremely complicated function and is of little use for computing $P(n)$. Sometimes, however, generating functions are useful in a computational sense, as we shall now see.

Suppose that $f(x)$ generates the Fibonacci numbers $F(n)$. That is,

$$f(x) = F(0) + F(1)x + F(2)x^2 + \cdots, \tag{3.7}$$

$$xf(x) = F(0)x + F(1)x^2 + F(2)x^3 \cdots, \tag{3.8}$$

and

$$x^2f(x) = F(0)x^2 + F(1)x^3 + F(2)x^4 + \cdots. \tag{3.9}$$

If we subtract (3.8) and (3.9) from (3.7) we obtain

$$(1 - x - x^2)f(x) = F(0) + F(1)x - F(0)x \tag{3.10}$$

because all further terms cancel out. [$F(n + 1) = F(n) + F(n - 1)$, remember?] But $F(0) = 1$, $F(1) = 1$, so we find

$$f(x) = \frac{1}{1 - x - x^2}. \tag{3.11}$$

Thus it was extremely easy to find the generating function for $F(n)$. The advantage to having a generating function lies in the fact that our knowledge of functions such as $f(x)$ is rather great, and we can often use algebraic and analytic methods to find the coefficients of f.

In this case, we can factor $1 - x - x^2$ as $(1 - \alpha x)(1 - \beta x)$. To find α and β we need the roots of $1 - x - x^2$, which are found to be $(-1 + \sqrt{5})/2$ and $(-1 - \sqrt{5})/2$, using the quadratic formula. Then $(1 - \alpha x) = 0$ yields $[1 - \alpha(-1 + \sqrt{5})/2] = 0$, so $\alpha = 2/(-1 + \sqrt{5}) = (1 + \sqrt{5})/2$. Similarly $\beta = (1 - \sqrt{5})/2$. We now write

$$\frac{1}{1 - x - x^2} = \frac{A}{1 - \alpha x} + \frac{B}{1 - \beta x} \tag{3.12}$$

and, using the techniques of partial fractions (borrowed from the calculus) we find $A = \alpha/(\alpha - \beta)$ and $B = -\beta/(\alpha - \beta)$. Thus

$$\begin{aligned} f(x) &= \frac{1}{1 - x - x^2} = \frac{A}{1 - \alpha x} + \frac{B}{1 - \beta x} \\ &= A\Sigma\alpha^n x^n + B\Sigma\beta^n x^n \end{aligned} \tag{3.13}$$

and $F(n)$ is the coefficient of x^n in $f(x)$, so that

$$\begin{aligned} F(n) &= A\alpha^n + B\beta^n \\ &= \frac{\alpha^{n+1}}{\alpha - \beta} - \frac{\beta^{n+1}}{\alpha - \beta} \\ &= \frac{\alpha^{n+1} - \beta^{n+1}}{\sqrt{5}} \end{aligned}$$

$$= \frac{1}{\sqrt{5}}\left[\left(\frac{1+\sqrt{5}}{2}\right)^{n+1} - \left(\frac{1-\sqrt{5}}{2}\right)^{n+1}\right]$$

$$= \frac{1}{2^n}\left[\binom{n+1}{1} + 5\binom{n+1}{3} + 5^2\binom{n+1}{5} + \cdots + 5^{m-1}\binom{n+1}{2m-1}\right]$$

if $n + 1 = 2m$ or $2m - 1$. (3.14)

Thus we have another formula for computing the Fibonacci numbers. For example,

$$F(0) = 1$$

$$F(1) = 1$$

$$F(2) = \tfrac{1}{4}[3 + 5] = 2$$

$$F(3) = \tfrac{1}{8}[4 + 4 \cdot 5] = 3$$

$$\cdots \qquad \cdots \qquad \cdots$$

$$F(10) = \tfrac{1}{2^{10}}\left[\binom{11}{1} + 5\binom{11}{3} + 5^2\binom{11}{5} + 5^3\binom{11}{7} + 5^4\binom{11}{9} + 5^5\binom{11}{11}\right] = 89$$

$$F(11) = \tfrac{1}{2^{11}}\left[\binom{12}{1} + 5\binom{12}{3} + 5^2\binom{12}{5} + 5^3\binom{12}{7} + 5^4\binom{12}{9} + 5^5\binom{12}{11}\right] = 144.$$

EXERCISES

1. Find the generating functions for the following sequences:
   (a) $1, 5, 5^2, 5^3, 5^4, 5^5, 5^6, 5^7, \ldots.$
   (b) $3, 2, 1, 0, 0, 0, 0, 0, \ldots.$
   (c) $1, -1, 1, -1, 1, -1, 1, -1, \ldots.$
   (d) $1, 0, 1, 0, 1, 0, 1, 0, \ldots.$
   (e) $0, 2, 0, 4, 0, 6, 0, 8, \ldots.$
   (f) $4, 8, 16, 32, 64, \ldots.$

2. Find an expression for the nth coefficient of the series generated by the following functions:
   (a) $\left(\frac{1}{1-x}\right)^3$
   (b) $\frac{1}{1-x}\,\frac{1}{1+x}$
   (c) $\frac{1}{1+4x}$
   (d) $\frac{2x}{(1-x)(1-2x)}$

[*Hint*: Use partial fractions.]

3. Find an expression for the nth coefficient of the series generated by the following functions:

(a) e^x

(b) $\sin x$

[*Hint*: Use derivatives.]

4. Find the generating function for the number of ways to make n cents change in pennies, nickels, dimes, and quarters.

5. Find the generating functions for partitions of n, restricted as follows:

(a) Into odd parts.

(b) Into even parts.

(c) Into parts, the largest of which is at most 4.

(d) Into parts, the largest of which is 4.

(e) Into at most 4 parts.

(f) Into exactly 4 parts.

(g) Into distinct parts.

6. Imitate the method used to find $F(n)$ in this section to find:

(a) $H(n)$, where $H(n) = 2H(n-1) - H(n-2)$, $H(0) = 0$, $H(1) = 1$.

(b) $G(n)$, where $G(n) = 3G(n-1) + 4G(n-2)$, $G(1) = G(2) = 1$.

3.6. RECURRENCE RELATIONS

Combinatorial problems are often solved by reducing them to problems involving a smaller number of objects. If a problem consists of a series of counting problems, depending on a parameter n, and there is sufficient regularity in the problems, the method of recurrence relations may be useful. The Fibonacci relation is an example. In general, given a sequence of numbers $\langle a_n \rangle$ a *recurrence relation* on $\langle a_n \rangle$ is a formula expressing a_n in terms of some a_i's, $i < n$, and a function of n. In particular, a linear recurrence of order k has the form

$$a_n = c_1 a_{n-1} + c_2 a_{n-2} + \cdots + c_k a_{n-k} + f(n).$$

The general theory is complex, and we shall only present some examples that show how such relations arise and how they are solved in a few simple cases.

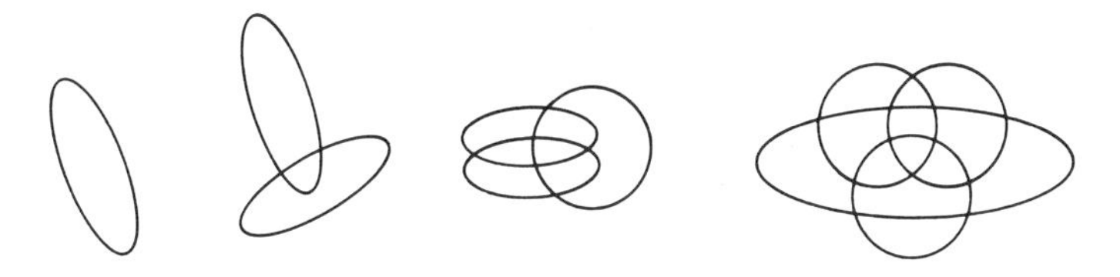

Figure 3.5

Example 12. If n ovals are drawn in the plane so that every pair of ovals meet in exactly two points and no three ovals meet at any point, how many regions are formed?

If the number of regions for n ovals is r_n, then Figure 3.5 shows that $r_1 = 2$, $r_2 = 4$, $r_3 = 8$, and $r_4 = 14$. Since diagramming is obviously not a method by which r_n can be found in general, we find a recurrence relation. The regions counted in r_n are of two types: those held over from r_{n-1} and those formed by splitting a region counted in r_{n-1}. How many regions are split and how many are left alone? Since the nth oval intersects each of $n - 1$ ovals in two points, the intersection points break it into $2(n - 1)$ segments, each of which splits an old region. Thus there are $2(n - 1)$ regions split and $r_{n-1} - 2(n - 1)$ regions left alone in adding the nth oval. This yields the recursion

$$r_n = [r_{n-1} - 2(n - 1)] + 2[2(n - 1)]$$

or

$$r_n = r_{n-1} + 2(n - 1). \tag{3.15}$$

To solve for r_n, we write

$$\begin{aligned} r_n &= r_{n-1} + 2(n - 1) \\ r_{n-1} &= r_{n-2} + 2(n - 2) \\ r_{n-2} &= r_{n-3} + 2(n - 3) \\ &\vdots \\ r_2 &= r_1 + 2. \end{aligned}$$

Adding and canceling we have

$$\begin{aligned} r_n &= r_1 + 2\{(n - 1) + (n - 2) + \cdots + 1\} \\ &= 2 + n(n - 1). \end{aligned}$$

Example 13. The tower of Hanoi puzzle (Figure 3.6) involves three pins and n discs; the discs are stacked in decreasing size on one pin. The object

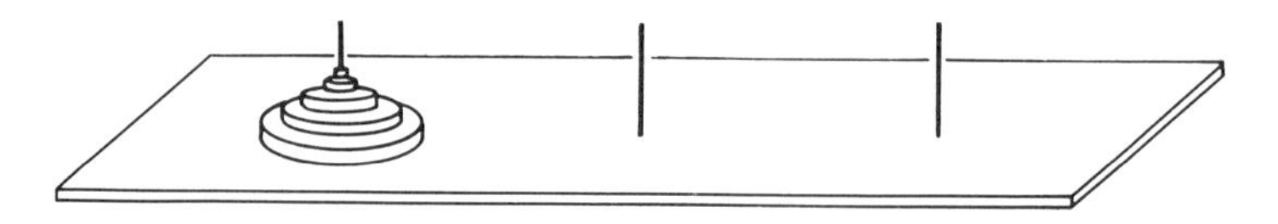

Figure 3.6

is to shift one disc at a time from one pin to another, never placing a larger disc on top of a smaller, and finally having the discs stacked in decreasing order on a different pin. If there are n discs how many shifts of single discs are required?

If $n = 1$ clearly one shift is sufficient. For $n = 2$ three moves are enough. If m_n denotes the number of shifts required to move a stack of n discs, then there is an easy recursion for m_n. Clearly the bottom disc can only be moved after the $n - 1$ discs above are moved to another pin, leaving the third pin free. This requires m_{n-1} shifts. Then the bottom disc is moved (one shift). Finally the stack of $n - 1$ discs must be placed on top of the largest one, again requiring m_{n-1} shifts. Thus we find

$$m_n = 2m_{n-1} + 1. \tag{3.16}$$

To solve for m_n, we proceed almost as before, with

$$\begin{aligned} m_n &= 2m_{n-1} + 1 \\ 2m_{n-1} &= 2^2 m_{n-2} + 2 \\ 2^2 m_{n-2} &= 2^3 m_{n-3} + 2^2 \\ &\vdots \\ 2^{n-2} m_2 &= 2^{n-1} m_1 + 2^{n-2}. \end{aligned}$$

Adding and canceling we find

$$\begin{aligned} m_n &= 2^{n-1} m_1 + (1 + 2 + 2^2 + \cdots + 2^{n-2}) \\ &= 2^{n-1} + (2^{n-1} - 1) \quad \text{since } m_1 = 1 \\ &= 2^n - 1. \end{aligned}$$

The next two examples involve a slightly more complicated type of recursion,

$$a_n = c_1 a_{n-1} + c_2 a_{n-2}. \tag{3.17}$$

Suppose that $a_n = r^n$ is a solution to (3.17). Then $a_{n-1} = r^{n-1}$ and $a_{n-2} = r^{n-2}$ so we have

$$r^n = c_1 r^{n-1} + c_2 r^{n-2}.$$

Thus r is a root of the equation

$$x^n - c_1 x^{n-1} - c_2 x^{n-2} = 0.$$

Since we are clearly not interested in the root $x = 0$, this can be reduced to

$$x^2 - c_1 x - c_2 = 0. \tag{3.18}$$

Equation (3.18) is called the *characteristic equation* of recurrence (3.17), and its roots are called *characteristic roots.* Each characteristic root r yields a solution $a_n = r^n$ to (3.17). Unfortunately, the solution r^n will probably not satisfy the initial conditions specifying particular values for a_0 and a_1 (or a_1 and a_2). If r_1 and r_2 are characteristic roots, so that r_1^n and r_2^n are solutions to (3.17), then $Ar_1^n + Br_2^n$ is also a solution for any choice of A and B. Choosing A and B carefully, we can find a solution that satisfies the specified initial conditions.

Example 14. A $2 \times n$ rectangle is to be paved with 1×2 and 2×2 blocks. In how many ways can that be done?

We assume that the rectangle is oriented so that ⊞ and ⊟ are different pavings of a 2×2 rectangle. Then if p_n is the number of pavings of a $2 \times n$ rectangle, clearly $p_1 = 1$ and $p_2 = 3$. A paving of a $2 \times n$ rectangle either begins with a paving of a $2 \times (n - 1)$ rectangle or it does not. Since there are p_{n-1} ways to pave a $2 \times (n - 1)$ rectangle, and only one way to continue such a rectangle to $2 \times n$, there are p_{n-1} pavings of the first type. If a paving does not begin with a $2 \times (n - 1)$ paving, it must begin with a $2 \times (n - 2)$ paving, and such a paving can be continued to a $2 \times n$ not containing a $2 \times (n - 1)$ in two ways (by adding ⊟ or ⊞). Thus there are $2p_{n-2}$ such pavings, and we find

$$p_n = p_{n-1} + 2p_{n-2}. \tag{3.19}$$

The characteristic equation of this recurrence is

$$x^2 - x - 2 = 0,$$

which has roots $x = 2$ and $x = -1$. Thus 2^n and $(-1)^n$ are possible solutions for (3.19), as is $A2^n + B(-1)^n$, for any choice of A and B. Choosing $n = 1$ and $n = 2$ we find

$$A(2) + B(-1) = a_1$$
$$A(2)^2 + B(-1)^2 = a_2$$

or

$$2A - B = 1$$
$$4A + B = 3.$$

Solving for A and B, we find $A = 2/3$, $B = 1/3$, so the solution to (3.19) that satisfies the specified initial conditions is

$$\begin{aligned} p_n &= (2/3)2^n + (1/3)(-1)^n \\ &= \frac{2^{n+1} + (-1)^n}{3}. \end{aligned}$$

Example 15. What is the determinant of the matrix

$$\begin{bmatrix} 2 & 1 & 0 & 0 & 0 & \dots & 0 & 0 \\ 1 & 2 & 1 & 0 & 0 & \dots & 0 & 0 \\ 0 & 1 & 2 & 1 & 0 & \dots & 0 & 0 \\ . & . & . & . & . & \dots & . & . \\ 0 & 0 & 0 & 0 & 0 & \dots & 1 & 0 \\ 0 & 0 & 0 & 0 & 0 & \dots & 2 & 1 \\ 0 & 0 & 0 & 0 & 0 & \dots & 1 & 2 \end{bmatrix},$$

which has a diagonal band (1, 2, 1) and is otherwise zero? If d_n is the determinant of an $n \times n$ matrix of this form, expanding the determinant by the terms of the first row yields

$$d_n = 2d_{n-1} - d_{n-2}. \tag{3.20}$$

This yields characteristic equation

$$x^2 - 2x + 1 = 0,$$

which has $x = 1$ as a repeated root. Our general solution is of the form $d_n = A(1)^n + B(1)^n$. Unfortunately, the repeated root makes it impossible to choose A and B to satisfy the initial conditions $d_1 = 2$ and $d_2 = 3$.

Note, however, that if r is a repeated root of $f(x) = 0$, then r is also a root of $f'(x) = 0$. In this example, 1 is a multiple root of $x^2 - 2x + 1 = 0$, or $x^n - 2x^{n-1} + x^{n-2} = 0$, and therefore also a root of $nx^{n-1} - 2(n-1)x^{n-2} + (n-2)x^{n-3} = 0$. Multiplying by x we find that $nx^n - 2(n-1)x^{n-1} + (n-2)x^{n-2} = 0$. Thus $d_n = n(1)^n$, as well as $(1)^n$, is a solution to the original recurrence relation (3.20). A general solution is then

$$d_n = An(1)^n + B(1)^n.$$

Since $d_1 = 2$ and $d_2 = 3$, we have

$$2 = A + B$$

$$3 = 2A + B.$$

Solving for A and B, we find $A = 1$, $B = 1$, so $d_n = n + 1$ is the solution.

The previous examples illustrate a general method for solving linear recurrence relations. Any linear recurrence relation

$$a_n = c_1 a_{n-1} + c_2 a_{n-2} + \cdots + c_k a_{n-k}$$

can be solved by finding the roots $r_1, r_2, \ldots, r_k$ of the characteristic equation

$$x^k - c_1 x^{k-1} - c_2 x^{k-2} - \cdots - c_k = 0.$$

If the roots r_i are distinct, the general solution is of the form

$$a_n = A_1(r_1)^n + A_2(r_2)^n + \cdots + A_k(r_k)^n$$

and the A_i's are chosen so that the initial values $a_1, a_2, \ldots, a_k$ are correct. If some root r_i is repeated m times then $(r_i)^n, n(r_i)^n, n^2(r_i)^n, \ldots, n^{m-1}(r_i)^n$ are used as distinct solutions.

If the recurrence has the form

$$a_n = c_1 a_{n-1} + c_2 a_{n-2} + \cdots + c_k a_{n-k} + f(n), \quad f(n) \neq 0, \qquad (3.21)$$

it is called *nonhomogeneous.* We solved some simple nonhomogeneous recurrences in Example 12 and 13 above. In general, the solution of non-homogeneous equations is beyond the scope of this text. The reader familiar with differential equations, however, will not be surprised that a solution to (3.21) is a sum of a solution to the corresponding homogeneous equation [obtained by ignoring $f(n)$] and a particular solution (obtained by ignoring the initial conditions). An example is presented in the exercises.

EXERCISES

1. A circle is drawn in the plane, and n straight lines are drawn so that each line intersects all of the other lines inside the circle, with no three lines meeting in a point. Into how many regions do those n lines divide the interior of the circle? The exterior?
2. How many of the 2^n binary n-sequences contain an even number of 0's?
3. How many of the 4^n quaternary n-sequences contain an even number of 0's?
4. Solve the recurrence $a_n = a_{n-1} + a_{n-2}, n \geqslant 2, a_0 = 1, a_1 = 3$.
5. Solve the recurrence $a_n = -6a_{n-1} - 12a_{n-2} - 8a_{n-3}, n \geqslant 3, a_0 = 1, a_1 = -2, a_2 = 8$.
6. Solve the recurrence $a_n = 3a_{n-2} - 2a_{n-3}, n \geqslant 3, a_0 = 1, a_1 = 0, a_2 = 0$.
7. Solve the recurrence $a_n + 2a_{n-1} = n + 3$, if $a_0 = 3$.
8. Find the determinant of the $n \times n$ matrix A with $a_{ii} = 2$, $a_{i,i-1} = a_{i,i+1} = -1$, all i, $a_{ij} = 0$ otherwise.
9. Suppose, in Fibonacci's rabbit problem, that each pair of rabbits, after producing two new pairs of rabbits, leave the rabbit colony. How many pairs of rabbits are in the colony in the middle of the nth month after starting with one pair of baby rabbits?
10. Suppose a code word is made up of A's, B's, and C's, and never has two A's in a row. Find a recursion for w_n, the number of n-letter code words.
11. At the age of one year a pair of Hungarian ground weasels produces two new pairs of weasels, and each year after that a pair produces six new pairs of weasels. Supposing that weasels never die and we begin with one newborn pair of weasels:
    (a) Write a recursion for $p(n)$, the number of weasel pairs after n years.
    (b) Solve that recursion.

3.7. REMARKS

In this chapter we have advanced far beyond the elementary methods of Chapter 1. Elaborating on the algebraic theme begun in Chapter 2, we have introduced three tools of modern enumeration—recurrence, generating functions, and the method of inclusion and exclusion. In each of these areas, we have presented only the most elementary aspects of a deep and rich theory.

Leonardo of Pisa (better known by the name Fibonacci) introduced recurrence relations to Europe in 1202 in an algebra text based on earlier Arab writings. The first use of generating functions in a combinatorial setting was De Moivre's work solving the Fibonacci recurrence more than 500 years later. Shortly thereafter, in the middle of the 18th century, Euler applied generating functions to partition problems, deriving many interesting results but no closed formulas for $P(n)$. Many methods for dealing with recurrences were developed in the study of difference equations (discrete analogues of differential equations). The principle of inclusion and exclusion is very old, but its first formal statement and proof appeared in works of da Silva and Sylvester about 100 years ago.

A most complete treatment of the topics of this chapter can be found in Riordan [1]. Treatments that are more in-depth than ours, and further counting methods, are also available in Liu [2] and Tucker [3].

[1] J. Riordan, *An Introduction to Combinatorial Theory*, Wiley, New York, 1958.

[2] C. L. Liu, *Introduction to Combinatorial Mathematics*, McGraw-Hill, New York, 1968.

[3] A. Tucker, *Applied Combinatorics*, Wiley, New York, 1980.

CHAPTER 4

Graphs

4.1. INTRODUCTION

Graphs, or linear graphs as they were once known, may be defined in several ways. We shall define a *graph* G to be a finite nonempty set $V = V(G)$ of p *vertices* together with a set X of q unordered pairs of distinct vertices of V. Each pair $x = \{u, v\}$ of vertices in X is an *edge* of G, and x is said to *join* u and v, while u and v are said to be *adjacent*. If G has p vertices and q edges, it is a *(p, q)-graph.* A graph H is a *subgraph* of G if $V(H) \subseteq V(G)$ and $X(H) \subseteq X(G)$, where, of course, any edge in H must have both its vertices in $X(H)$

The set-theoretic definition we have given has the virtue of precision, but graphs are often defined and represented in at least two other ways. The terminology we have used suggests a geometric representation, such as that shown in Figure 4.1(a). This graph has four vertices and five edges. In Figure 4.1(b) this same graph is represented as a matrix, known as the *adjacency matrix* of the graph. The ith column and row of the matrix are identified

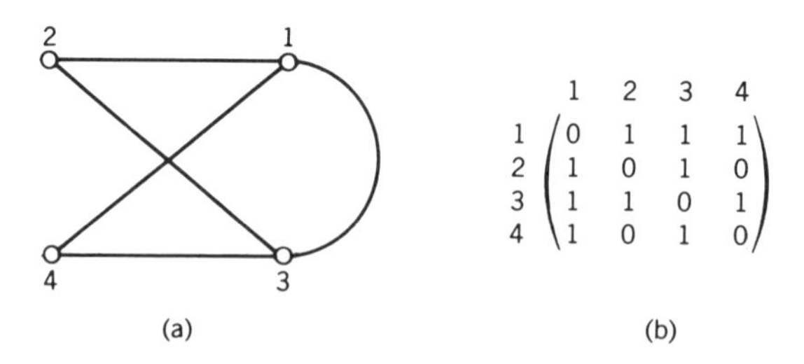

Figure 4.1

with the ith vertex of the graph, and $a_{ij} = 1$ whenever vertex i is adjacent to vertex j. Each of these representations introduces extraneous factors that must be examined. In the diagram, two edges are special because they cross, and one edge is special because it is curved. Since we have defined edges only in terms of the vertices that they join, these special properties of the diagram we have chosen should be ignored. Similarly, the matrix representation assumes that the vertices are ordered in some way, with a first vertex, a second, and so on. This ordering is not relevant, however, to what graph we are representing and should be suppressed.

We call two graphs *isomorphic* if there is a one-to-one correspondence between their vertex sets that preserves adjacency. Thus, two vertices are joined by an edge in one graph if and only if the corresponding vertices are joined by an edge in the other. For the pictorial representation of a graph, this means that the two diagrams will coincide if the vertices are moved to suitable positions and the edges distorted to similar shapes. It also implies that the matrices of the two graphs will coincide if the rows and columns are reordered in the proper way. In other words, the graphs are isomorphic if and only if the matrices are similar.

The theory of graphs is a large and rapidly growing area of mathematics, which arose out of investigations of various phenomena in the real world. Networks of various kinds, including highways, sewers, communication nets, and electrical networks can all be represented by graphs. Chemical compounds are graphs, with atoms for vertices and chemical bonds for edges. Flow diagrams for computers are graphs. Each application of graphs gives rise to numerous questions, which when answered give information not only about the original application but many others as well.

As an example of this, let us examine a way that graphs arise in set theory. Given any set S and a finite collection C of subsets of S, the *intersection graph* $I(C)$ has as vertices the members of C, and two vertices are adjacent if and only if the corresponding subsets have a nonempty intersection. If, for example, $S = \{1, 2, 3, 4, 5, 6\}$ and $C = \{\{1, 2\}, \{2, 4\}, \{1, 2, 3\}, \{3, 4, 5\}, \{5, 6\}\}$ then $I(C)$ is as shown in Figure 4.2(a).

Although the collection C must be finite, and the set S of this example was also finite, S is infinite in many of the most interesting examples. If S is the set of all real numbers arranged on the number line, and C is a collection of intervals on S, $I(C)$ is called an *interval graph*. Although every graph is an intersection graph, not every graph is an interval graph (see the exercises). The graph of Figure 4.2(a) is the interval graph of the intervals shown in exploded form in Figure 4.2(b).

Interval graphs were first used in biology to study the structure of genes. To test the hypothesis that genes are linear rather than circular, branched, or even more complicated, experiments were performed to discover whether

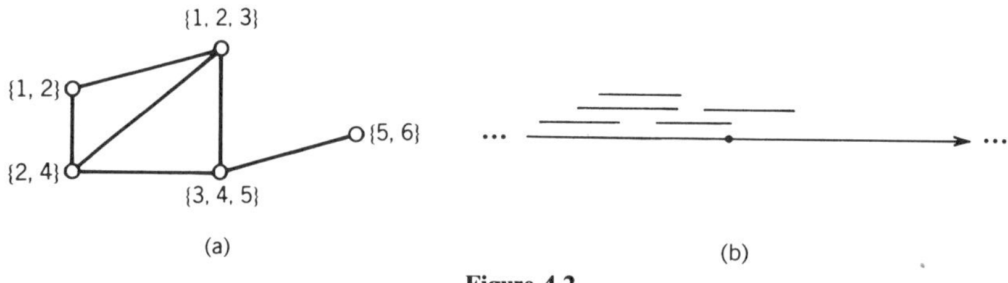

Figure 4.2

or not two fragments of a gene (called mutants) overlapped. Such information about many pairs of mutants was collected, and then the intersection graph of the mutants was tested to see if it was an interval graph. It was, and that tended to support the hypothesis of a linear arrangement.

Some years later the theory of interval graphs was applied in archaeology. Prehistoric graves contain various types of relics, which help to date the finds. Each type of relic was produced during a continuous interval of time. It seems reasonable to assume that relics found together were produced during periods of time that overlap, and if two types never appear together they were produced at different times. Thus, the information from a large number of excavations gives information on the adjacencies of various time intervals, and the theory of interval graphs is used to construct all possible time-interval models.

Note that these applications are quite different. In one the intervals are physical linkages in genes, in the other they are intervals of time. Only when we have created a model can we see the similarity of the problems.

The power of graph theory, and mathematics in general, is to isolate an essential element of structure in a complex situation so that it can be analyzed more fully and exactly.

EXERCISES

Note. All graphs are finite, no loops or multiple edges are allowed.

1. Draw all graphs with four vertices.

2. Show that the following graphs are isomorphic.

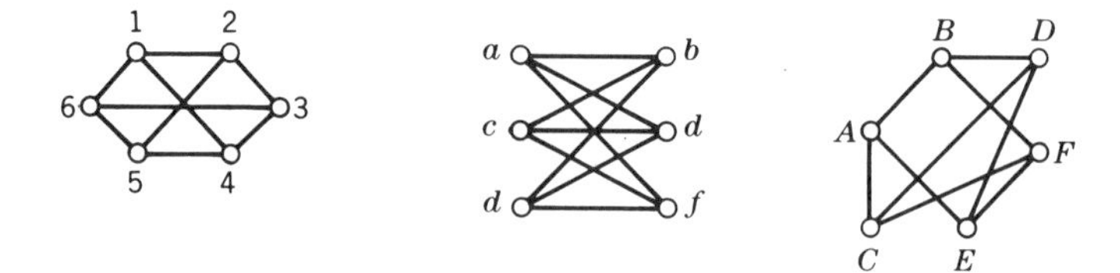

3. Show that one of the following graphs is not a subgraph of the graph in exercise 2 and the other one is a subgraph of that graph.

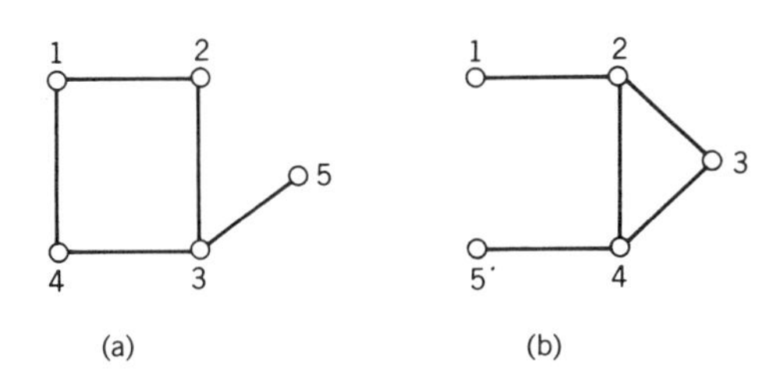

4. For an arbitrary graph G:

(a) What can be said about the entries a_{ii} of the adjacency matrix A?

(b) How are a_{ij} and a_{ji} related?

(c) What matrices are adjacency matrices of some graph?

5. (a) Write the adjacency matrix A of the graph (a) in exercise 3.

(b) Find A^2.

(c) Interpret the entries of the matrix A^2 in geometric terms.

6. Draw the intersection graph of the following collection of sets: $C = \{\{1, 3, 4\}, \{2, 3, 4\}, \{2, 3, 5\}, \{2, 4\}, \{1\}, \{5, 6\}, \{1, 6\}\}$.

7. (a) Prove that every graph is the intersection graph of some collection of sets. [*Hint*: An edge can be considered as a set of vertices.]

(b) Give an example of a graph on four vertices that is not an interval graph and show that it is not.

8. Suppose a, b, c, d, e, and f represent six kinds of relics. Draw a graph of adjacencies and use it to give at least two possible sets of chronological intervals that explain the following sets of grave contents: $\{e, f\}$, $\{a, c, e\}$, $\{b, e, f\}$, $\{b, d, f\}$, $\{a, e\}$, $\{b, d\}$.

4.2. BASIC CONCEPTS

If v is a vertex of a graph, then the *degree* of v, $d(v)$, is the number of edges incident with (i.e., containing) v. There is a simple relationship between the degrees of the vertices of G and the number of edges in G.

Theorem 4.1. If the vertices of G have degrees $d_1, d_2, \ldots, d_p$, and G has q edges, then $\Sigma d_i = 2q$.

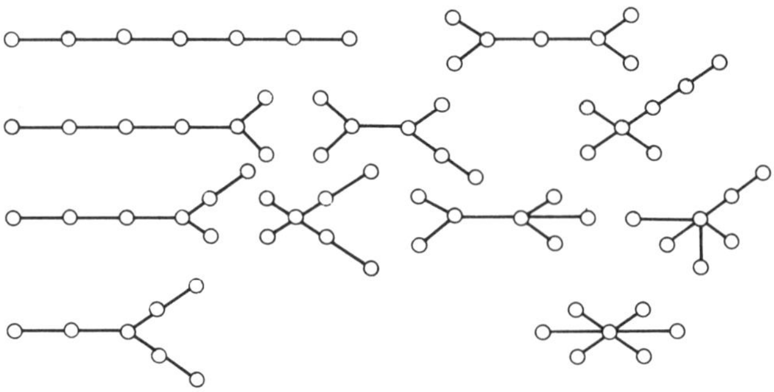

Figure 4.3

Proof. The left-hand side counts each edge once for each of the two vertices that contain it. Thus every edge is counted twice.

A *walk* in a graph or a multigraph is an alternating sequence of vertices and edges, $v_0, x_1, v_1, x_2, \ldots, v_{n-1}, x_n, v_n$ beginning and ending with vertices, in which each edge is incident with the vertices immediately preceding and following it. This walk *joins* v_0 and v_n and may be called a (v_0, v_n) walk. It is *closed* if $v_0 = v_n$ and *open* otherwise. It is a *trail* if the edges are distinct and a *path* if all the vertices are distinct. If $v_0 = v_n$, but all other vertices are distinct, it is a *cycle.*

A graph is *connected* if every pair of vertices are joined by a path. If a graph is not connected, it is called *disconnected* and its maximal connected subgraphs are called *components.*

A simple and important type of graph is called a tree. A *tree* is a connected graph containing no cycles. The trees with seven vertices are shown in Figure 4.3.

Three other special types of graphs will also be of interest to us. A graph is *regular* of degree d if every vertex has degree d. The graph on n vertices in which every vertex is joined to every other vertex is called the *complete n-graph* and denoted K_n. Finally, if the vertices of a graph G can be partitioned into two sets, V_1 and V_2, so that every edge joins a vertex of V_1 to a vertex of V_2, G is called *bipartite.* If every vertex of V_1 is joined to every vertex of V_2 then G is *complete bipartite* and is called $K_{m,n}$, where $m = |V_1|$ and $n = |V_2|$.

Bipartite graphs turn up frequently in applications. Consider, for example, the following bracing problem. A rectangular grid constructed of rigid struts can flex in many ways if no diagonal braces are included. Even if some braces are present, flexing may be possible. See Figures 4.4(a) and (b). Obviously braces in every space will prevent flexing, but more efficient solutions should be possible. Which bracings make a grid rigid? What is the smallest number of

Figure 4.4

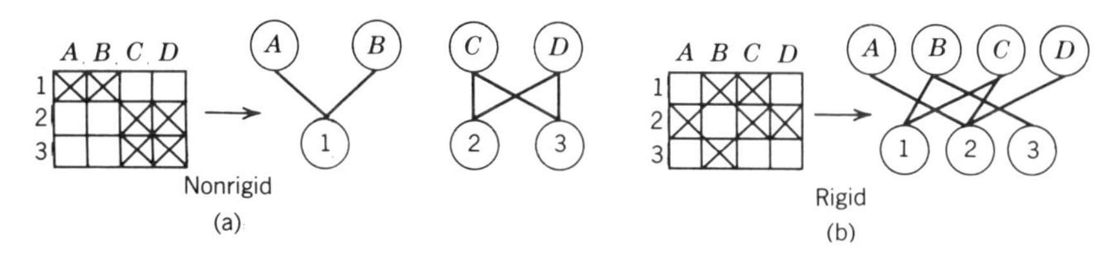

Figure 4.5

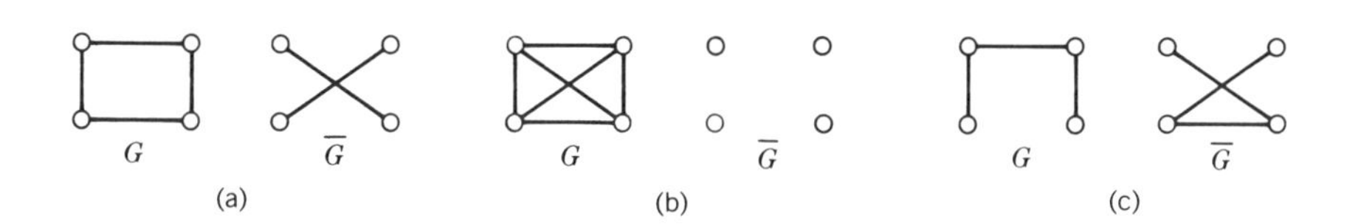

Figure 4.6

spaces that need to be braced on an $m \times n$ grid? To solve these problems we introduce a bipartite graph B whose two parts V_1 and V_2 are the columns and rows of the grid. A row and column are adjacent in B if the space at their intersection is braced. See Figure 4.5. Then it can be proved that a braced grid is rigid if and only if the corresponding bipartite graph is connected (see the exercises).

The complement of a graph G is everything that G is not. That is, the *complement* $\bar{G}$ has the same set of vertices as G, but two vertices are adjacent in $\bar{G}$ if and only if they are not adjacent in G. We display some examples in Figure 4.6. Note that the graphs of Figure 4.6(c) are isomorphic. A graph that is isomorphic to its complement is called *self-complementary*.

The definition of $\bar{G}$ implies immediately that the complement of the complement of G is G, or $\bar{\bar{G}} = G$. The following theorem is slightly less obvious.

Theorem 4.2. If G is disconnected then $\bar{G}$ is connected.

Proof. Suppose G is not connected. If u and v are any two vertices of G, we must find a path in $\bar{G}$ from u and v. If u and v are in different components of G, then they cannot be adjacent in G, and hence must be adjacent in $\bar{G}$, so there is a path of length 1 connecting them in $\bar{G}$. If u and v are in the same component of G, they may or may not be adjacent in G. But since G has more than one component, there must be a vertex, say w, in another component of G, which is adjacent to neither u nor v in G. Thus w is adjacent to both u and v in $\bar{G}$, and so there is a path of length 2 connecting u and v in $\bar{G}$. Thus we have shown that u and v, an arbitrary pair of vertices, are joined by a path in $\bar{G}$. So $\bar{G}$ is connected.

The following corollary is immediate.

Corollary 4.1. Every self-complementary graph is connected.

EXERCISES

1. For each of the following sequences of numbers, either find a graph that has the numbers as the degrees of its vertices or show there is no such graph.
   (a) 4, 4, 4, 3, 3
   (b) 4, 4, 3, 3, 3,
   (c) 4, 3, 2, 1
   (d) 5, 5, 4, 3, 2, 1
2. Prove that in any graph the number of vertices of odd degree is even.
3. Show that if 37 people are at a party, at least one of them shakes hands with an even number of people there.
4. Construct a graph that is regular of degree 3 and contains no cycles of length 3.
5. Prove that if every vertex of G has degree at least 2, G must contain a cycle.
6. Show that a graph G is a tree if every pair of vertices of G are joined by a unique path.
7. Find the number of edges in K_n and in $K_{m,n}$ for arbitrary m and n.
8. Find the number of edges in an r-regular graph on p vertices.
9. Prove that a bipartite graph cannot contain any cycles of odd length.
10. Prove that in a connected graph any two longest paths must have a vertex in common.

11. Prove that every rigid bracing of an $m \times n$ grid system has at least $m + n - 1$ braced rectangles.

12. Prove that a bracing is rigid if and only if the corresponding bigraph G is connected. [*Hints*: To show connected implies rigid, reduce G to a tree, delete an endvertex, and reduce to smaller case. To show rigid implies connected, assume a smallest counterexample and derive a contradiction.]

13. Show that there is no self-complementary graph on six vertices.

14. Show that every self-complementary graph has $4n$ or $4n + 1$ vertices, for some integer n.

4.3. EULERIAN TRAILS

Graph theory, as well as topology in general, began in 1736 when the Swiss mathematician Euler solved a problem that had plagued the people of the city of Königsberg for many years. It seems there were two islands in the Pregel River, connected to each other and the banks of the river by seven bridges as shown in Figure 4.7. The good people of Königsberg amused themselves on pleasant Sunday afternoons by strolling about the town, attempting to find a route that would take them over each bridge exactly once and return them to their starting place. Euler showed that no such route existed by posing and then solving a more general problem. He replaced each land area by a vertex and each bridge by an appropriate edge. The diagram thus formed is shown in Figure 4.8. This is a *multigraph* because it contains several edges joining some pairs of vertices. The land areas on the right and left banks are represented by the vertices 1 and 2, and the two islands are represented by the vertices 3 and 4.

The problem posed calls for a walk on this multigraph that traverses each edge exactly once and returns to its starting point. Such a walk is called a

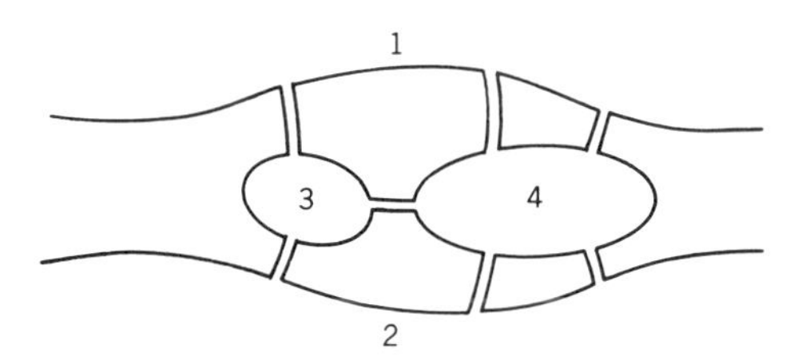

Figure 4.7

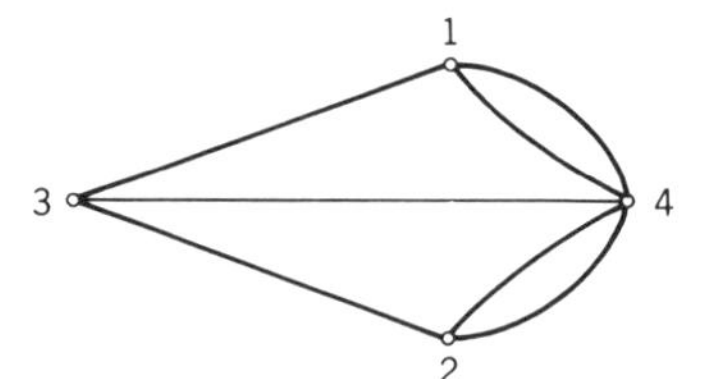

Figure 4.8

eulerian trail, and a multigraph with such a trail is called *eulerian*. Euler found the following criterion for a multigraph to be eulerian.

Theorem 4.3. A connected multigraph G is eulerian if and only if every vertex of G has even degree.

Proof. Let T be a eulerian trail of G. Then each occurrence of a given vertex in T contributes 2 to the degree of the vertex. Since every edge lies on T exactly once, the degree of the vertex is a sum of 2's, so it is even.

On the other hand, suppose G is connected and every vertex has even degree. We have to find a closed eulerian trail in G. Select any vertex of G and form a trail T starting at that vertex as follows. Each time a vertex is reached, select any edge that has not yet been used and continue to another vertex. Because each degree is even, for every edge into a vertex there is an edge out. The only exception to this rule is our starting vertex, where we have used an edge out without a corresponding edge in. Thus if we reach a vertex that has no more edges on which to continue T, we must have returned to our starting vertex. If, when this happens, T contains all edges of G, we have the required eulerian trail. Otherwise, we have a closed trail in G, any vertex of which may be taken as a starting point. Since G is connected, some vertex on the trail T is incident with an edge not in T. Using that vertex as a new starting point we can form a new trail T' that contains no edges from T and must return to its starting vertex since in $G-T$ each vertex still has an even degre. Then T and T' taken together are also a closed trail. If every edge of G is in one of them, we have a eulerian trail. If not, we can continue to form new closed trails and add them to what we have until we have a eulerian trail, which proves the theorem.

Note that since the theorem applies equally well to multigraphs as to graphs, it solves the Königsberg bridge problem. In fact, it also applies to even more general structures. In particular, it is clearly not a problem if we allow edges (called loops) that join a vertex to itself, as long as we agree that each loop contributes 2 to the degree of its vertex. In fact, the argument we have used applies equally well to graphs whose edges are directed. A *digraph* consists of a finite nonempty set V of *vertices* together with a set X of ordered pairs of vertices called *edges*. The *edge* $< u,v >$ goes *from u to v*, and u is *adjacent to v*, while v is *adjacent from u*. All of the concepts of undirected graphs may be generalized in a natural way to digraphs, with small arrows to indicate direction in pictorial representations, and walks, trails, paths, and cycles following edges in the correct direction. A digraph is *strongly connected* if every two vertices are joined by (directed) paths in both directions. The idea of degree must be generalized to count separately the edges into a vertex (the *indegree*) and edges out of a vertex (the *outdegree*). The analog of Theorem 4.3 is the following theorem.

Theorem 4.4. A strongly connected digraph D is eulerian if and only if every vertex of D has equal indegree and outdegree.

One application of eulerian digraphs has turned up in many ways, including construction of mechanical encoding machines. If a drum is constructed with eight segments along the boundary, four electrically live (1), and four dead (0) then a rotation of the drum against three contacts can give every possible combination of on and off patterns. The drum is shown in Figure 4.9 reading 1 0 1. Note that as the drum is rotated it will read off 1 0 1, 0 1 0, 1 0 0, 0 0 0, 0 0 1, 0 1 1, 1 1 1, 1 1 0, and then return to 1 0 1, giving each of the eight possible patterns exactly once. Is it possible to construct such a drum with 16 segments giving all 16 4-digit binary readings? What about 32? 64? 2^n for any n?

We can solve these problems if we realize that patterns of length k that are adjacent on the drum must overlap, so that the last $k - 1$ places of one are the same as the first $k - 1$ places of the next. Thus, we form a directed graph with all $k - 1$ sequences as vertices. The edges are labeled with k sequences and are directed from the $k - 1$ sequence that begins the edge to the $k - 1$ sequence that ends it. The graph corresponding to the drum of Figure 4.9 is shown in Figure 4.10. (Note that this digraph has loops, so it is actually a pseudodi-

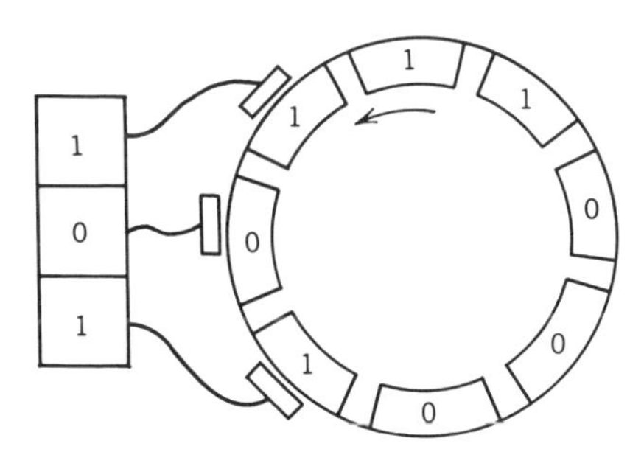

Figure 4.9

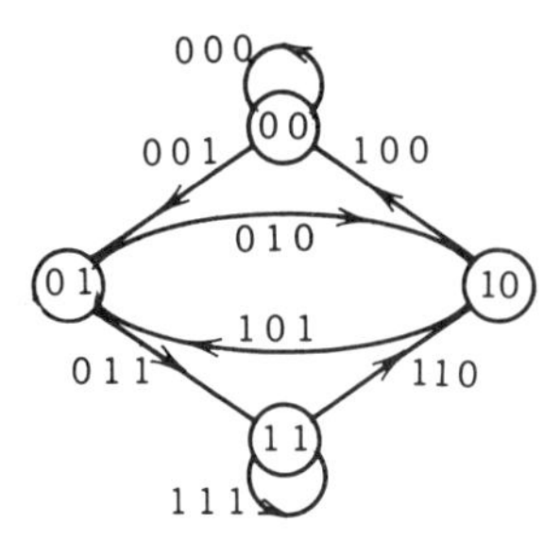

Figure 4.10

graph.) A sequence of eight edges that form a directed eulerian trail will determine a way to arrange sequences around a drum. How do we know such a trail exists? It is not hard to see that every vertex will have indegree and outdegree 2, so by Theorem 4.4 we have a eulerian trail. This generalizes to longer sequences (see the exercises), so in fact a wheel of length 2^n exists for every n.

EXERCISES

1. Which of the following graphs is eulerian? Find a eulerian trail for those that are.

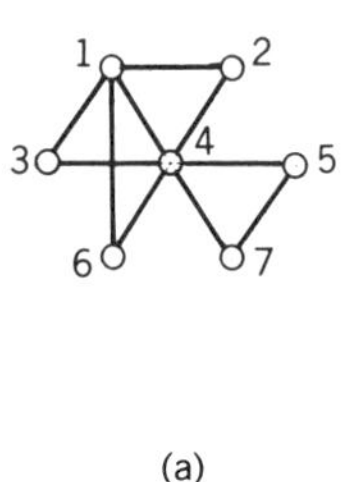

(a)

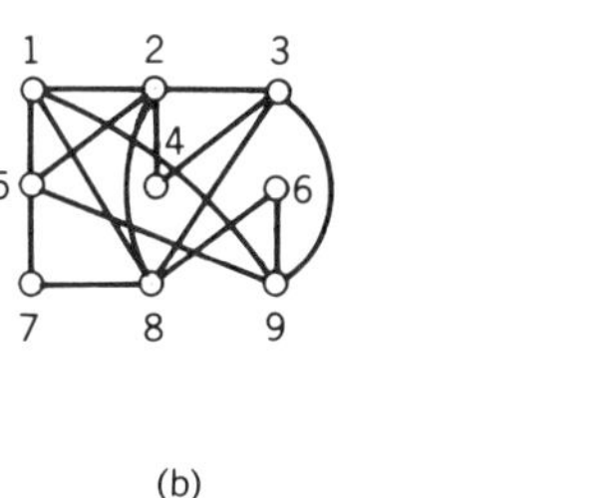

(b)

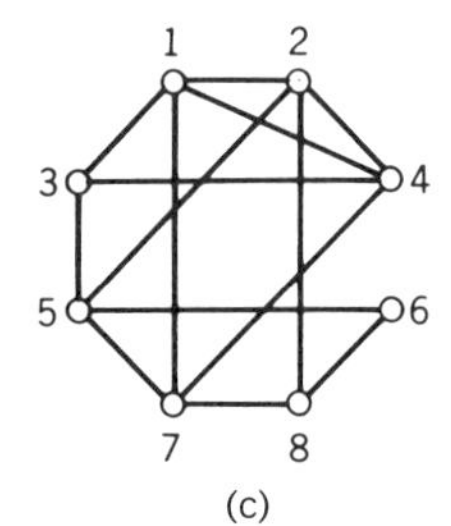

(c)

2. Is there a trail on each of the maps on the next page that crosses each bridge exactly once and ends at its starting place? If the answer is in the affirmative list the bridges in the order crossed.

3. A graph is *semi-eulerian* if it contains a trail (possibly open) that includes all edges. Prove that a graph is semi-eulerian if and only if it contains at most two vertices of odd degree.

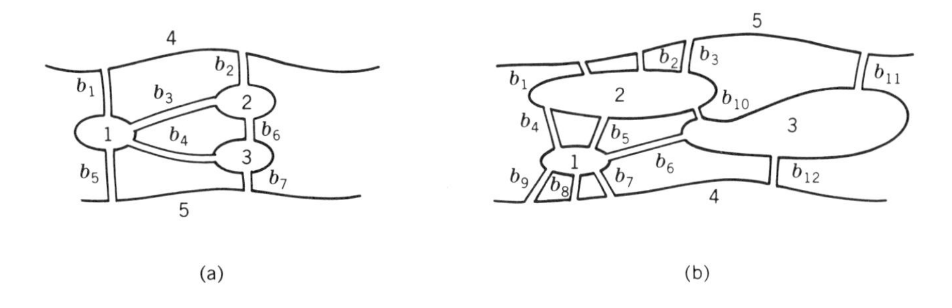

4. For each of the following figures, draw a continuous curve with your pencil that cuts each line exactly once and ends where it began, or else show that it cannot be done. (A line here means a line segment between junctions of lines or corners.)

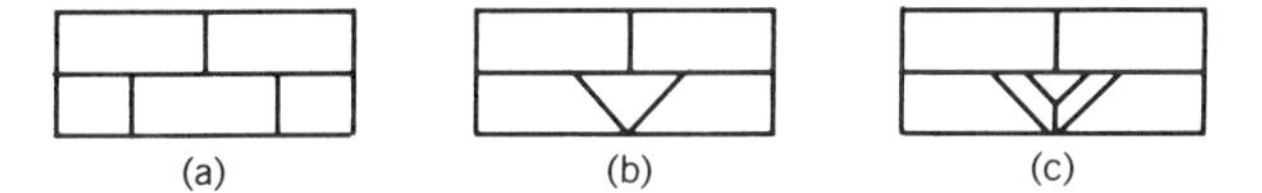

5. A set of double-six dominoes contains one domino of each type from blank–blank to six–six.
   (a) How many dominoes are in a set?
   (b) Dominoes may be placed end-to-end if the corresponding ends match. Is it possible to arrange the full set of double-six dominoes in a circle, obeying that rule?
   (c) Answer part (b) for double-nine dominoes.
6. Use a suitable directed graph to construct a code wheel for 4-digit binary sequences.
7. Prove that a code wheel reading off all q^n n-digit q-ary codes exists for every q and n, positive integers. (Binary wheels have $q = 2$.)

4.4. HAMILTONIAN GRAPHS

Another type of walk on a graph originated with Sir William Hamilton when, in 1859, he invented a game called "Around the World." The object of the game, pictured in Figure 4.11, was to start in London and take a trip around the world, visiting each city exactly once and, of course, traveling along the lines shown.

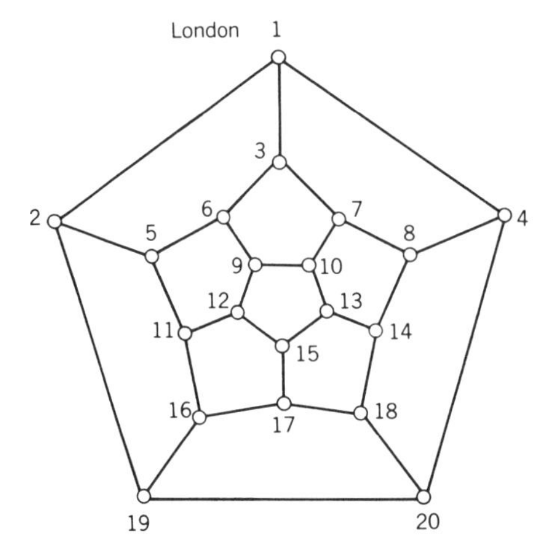

Figure 4.11

Hamilton sold his idea to a maker of games for 25 guineas; he proved a more astute businessman than gamesman, since the game was not a commercial success. Our point of view toward this puzzle is that of searching for a cycle that contains all vertices of the graph. A graph with such a spanning cycle is called *hamiltonian.* The graph used for this game is hamiltonian, so that there is a solution. For example, a spanning cycle is 1, 2, 19, 20, 4, 8, 14, 18, 17, 16, 11, 5, 6, 9, 12, 15, 13, 10, 7, 3, 1.

There is no simple method for determining whether or not a graph is hamiltonian. We will present some of the partial results that have been found.

A graph is hamiltonian if its degrees are large enough, as shown in the next theorem.

Theorem 4.5. If a graph G has $p \geqslant 3$ vertices, each of degree at least $p/2$, then G is hamiltonian.

Proof. We suppose that the theorem is false and derive a contradiction, which shows that our assumption is incorrect. If the theorem is false, there is some graph as specified in the hypothesis that is not hamiltonian. Let G be such a counterexample to the theorem, chosen so that no graph on p vertices with more edges than G is also a counterexample. We chose G in that special way because we may then assume that every pair of nonadjacent vertices are joined by a path containing all vertices of G. This can be seen as follows: Let the vertices v_0 and v be nonadjacent in the graph G chosen. Since G is the largest counterexample, if we add the edge $\{v_0, v\}$ to G, then we must get a hamiltonian

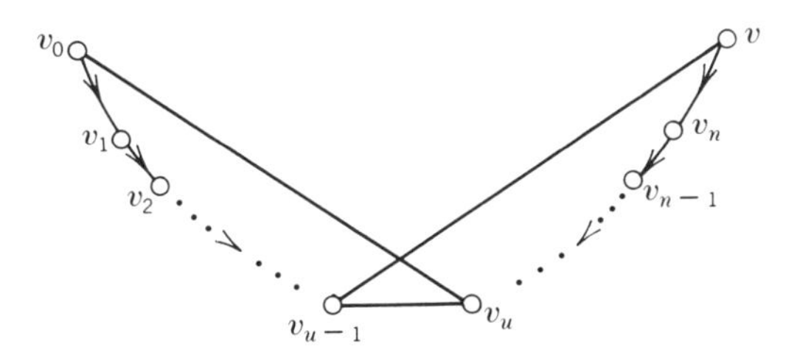

Figure 4.12

graph G_1. Hence, v_0 and v are on a spanning cycle. Removal of the edge $\{v_0, v\}$ gives a spanning path in G from v_0 to v along the cycle. Thus any two nonadjacent vertices in G are joined by a spanning path. We now proceed to exploit this property.

Say v_0 and v are two nonadjacent vertices of G, and $v_0v_1v_2 \cdots v_nv$ is a spanning path joining them. Let k be the degree of v_0. Then $k \geqslant p/2$. Let S be the set of vertices adjacent to v_0. Then $|S| = k$. We shall show that if $v_u \in S$, then v_{u-1} cannot be adjacent to v. If possible let v be adjacent to v_{u-1}. Then we have the hamiltonian cycle $v_0, v_1, v_2, \ldots, v_{u-1}, v, v_n, v_{n-1}, \ldots, v_u, v_0$, as in Figure 4.12. Since we are assuming that G is not hamiltonian, this is not possible so v is not adjacent to the vertex v_{u-1} if $v_u \in S$. Hence v is nonadjacent to at least k of the $p - 1$ vertices other than v. Thus the degree of v is at most $(p - 1) - k \leqslant p - 1 - (p/2) = (p/2) - 1$. This is a contradiction. Thus G must be hamiltonian, and the theorem is proved.

Note that although the conditions of this theorem are sufficient to assure that a graph is hamiltonian, they certainly are not necessary. For example, a 5-cycle is certainly hamiltonian, but does not have sufficiently high degrees to satisfy the hypotheses of the theorem.

In order to present a necessary condition and another sufficient condition for a graph to be hamiltonian, we need to examine the connectivity of a graph. Recall that a graph is connected if any two vertices are joined by a path. In order to describe quantitatively just how strongly a graph is connected, we define the *connectivity* of G, $k(G)$ to be the minimum number of vertices that must be removed from G in order to produce a disconnected or trivial (i.e., one vertex) graph. Of course, when a vertex is removed, all edges incident with it are also removed. The mention of the trivial graph is made necessary since a complete graph can never be made into a disconnected graph by removing vertices. A graph G is called *n-connected* if $k(G) \geqslant n$. A single vertex whose removal disconnects the graph is called a *cut-vertex*.

Besides connectivity, the statement of our theorem involves a special kind of graph. A *theta-graph* consists of two vertices of degree 3, joined by three disjoint paths of length at least 2. We now state a theorem that combines a necessary and a sufficient condition for possession of a hamiltonian cycle.

Theorem 4.6. Every hamiltonian graph is 2-connected and every non-hamiltonian 2-connected graph contains a theta-graph.

Proof. If G is hamiltonian, then clearly it is connected. Furthermore, since any vertex is on the spanning cycle its removal cannot result in a disconnected graph. Hence G is 2-connected.

Suppose G is 2-connected. We shall show that it contains a cycle. First we note that each vertex must have degree at least 2. If not there is a vertex v_0 of degree. 1. Removal of the vertex adjacent to v_0 makes the graph disconnected. which is a contradiction. Let v be any vertex and let a and b be adjacent to it. The removal of v does not disconnect the graph. Hence there must be a path P from a to b not containing the edges $\{v, a\}$ and $\{b, v\}$. Adding these edges to P we have a cycle.

If G is not hamiltonian let C be the longest cycle. There must exist a vertex or vertices not on C (otherwise G would be hamiltonian). Since G is connected there must exist a vertex v not on C, adjacent to some vertex u of C. Since moreover G is 2-connected there must be a path from v to another vertex w on C; otherwise, the removal of u would break the graph into two or more components. Since C is chosen as the maximal cycle, u and w cannot be adjacent on C; otherwise, the replacement of the edge $\{u, w\}$ by the path from u to w through v would yield a longer cycle. Thus we have a theta-graph as shown in Figure 4.13.

Theorems 4.5 and 4.6 can be improved, but all known results on hamiltonian graphs are very far from being characterizations. No one knows a "good" criterion for hamiltonicity in graphs.

For certain directed graphs, more can be said. A digraph is hamiltonian if it has a (directed) path through all of its vertices. It has a *hamiltonian path* if it has

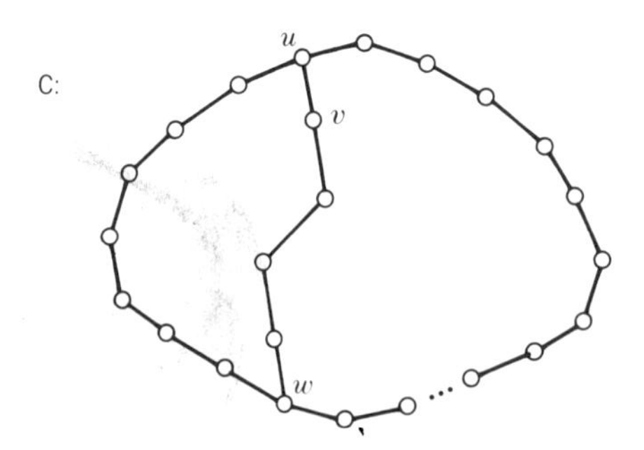

Figure 4.13

a (directed) path through all of its vertices. If the edges of K_n are directed, we obtain a *tournament*, so called because it can be viewed as the win–loss record of a round-robin tournament among players who correspond to the vertices. Although there is only one complete graph K_n on n vertices, there are many tournaments T_n on the same vertices, because the edges can be directed in various ways.

Theorem 4.7. A directed complete graph T_n always contains a hamiltonian path.

Proof. Let there be a path of length $k - 1$ in the digraph which meets the sequence of vertices $(v_1, v_2, \ldots, v_k)$. Let v_0 be a vertex that is not included in this path. If there is an arc $\langle v_0, v_1 \rangle$ in the digraph, we can augment the original path by adding the arc $\langle v_0, v_1 \rangle$ to the path so that the vertex v_0 will be included in the augmented path. If, on the other hand, there is no arc from v_0 to v_1, then there must be an arc $\langle v_1, v_0 \rangle$ in the digraph. Suppose that $\langle v_0, v_2 \rangle$ is also an arc in the digraph. We can replace the arc $\langle v_1, v_2 \rangle$ in the original path with the two arcs $\langle v_1, v_0 \rangle$ and $\langle v_0, v_2 \rangle$ so that the vertex v_0 will be included in the augmented path. On the other hand, if there is no arc from v_0 to v_2, then there must be an arc $\langle v_2, v_0 \rangle$ in the path and we can repeat the argument.

Eventually, if we find that it is not possible to include the vertex v_0 in any augmented path by replacing an arc $\langle v_i, v_{i+1} \rangle$ in the original path with two arcs $\langle v_i, v_0 \rangle$ and $\langle v_0, v_{i+1} \rangle$ with $1 \leqslant i < k - 1$, we conclude that there must be an arc $\langle v_k, v_0 \rangle$ in the graph. We can, therefore, augment the original path by adding to it the arc $\langle v_k, v_0 \rangle$ so that the vertex v_0 will be included in the augmented path. Repeating the argument, we can include all vertices in a hamiltonian path.

As an application of Theorem 4.7, consider the following cleaning problem. The head of a janitorial service has contracted to straighten and clean all n of the offices in a large office building. The work in office i is conveniently divisible into a straightening component, which takes s_i time, and a cleaning component, which takes c_i time. The values s_i and c_i depend on various factors, including the size of the office, but cleaning is generally the harder job, so it is reasonable to assume that for any i and j either $c_i \geqslant s_j$ or $c_j \geqslant s_i$. Is it possible for the head janitor to order the offices so that after he straightens each office his assistant cleans it, and his assistant never needs to wait for an office to clean? (Of course, the assistant will have to wait until there is one office straightened before he starts cleaning.)

To see that such a schedule is possible, we define a graph with a vertex for each office, and put an arc from i to j if $c_i \geqslant s_j$. Our assumption assures us that each pair of offices are joined at least one way, so we have (at least) a directed complete graph. Note that a hamiltonian path in the graph will specify an ordering of offices which always keeps the assistant busy because he starts out behind and takes at least as long to clean office i as the boss does to straighten office j, the next one on the schedule. Thus, Theorem 4.7 assures us that there is a schedule as required.

EXERCISES

1. Find a route "around the world."

2. Find whether or not each of the following graphs is hamiltonian. For those that are, display a spanning cycle. For those that are not, give an argument that proves they are not.

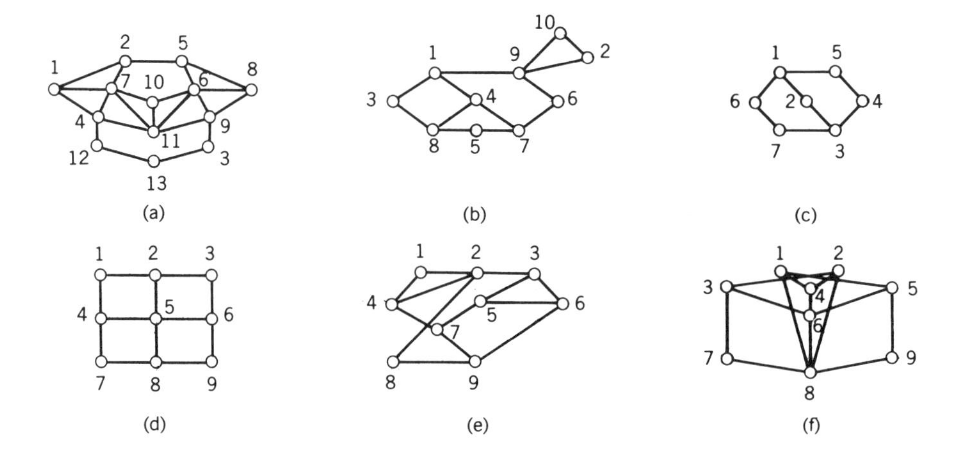

3. Show that any graph with $p \geqslant 3$ vertices is hamiltonian if for every pair u, v of nonadjacent vertices $\deg(u) + \deg(v) \geqslant p$.

4. Find the connectivity of the following graphs.

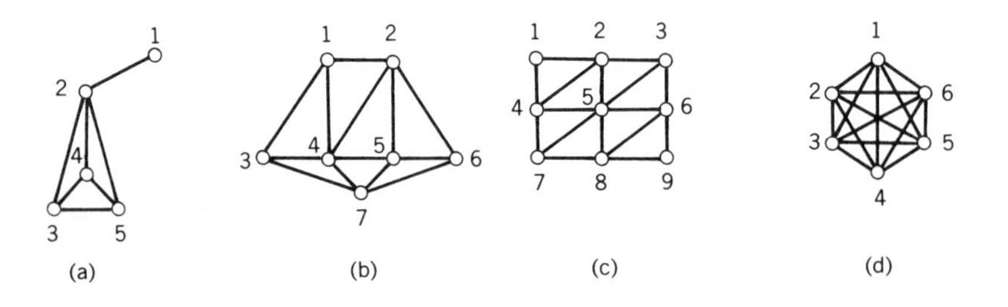

5. Show that a bipartite graph that is hamiltonian has an even number of vertices.

6. (a) Find a nonhamiltonian graph with four vertices and as many edges as possible.

(b) Do the same thing for 5- and 6-vertex graphs.

(c) Find a general example that works for n vertices, and thus find an upper bound on the number of edges in a nonhamiltonian p-graph.

7. Show that if v is a cutvertex of G, then v is not a cutvertex of $\bar{G}$.

8. Consider the lines of an 8×8 chessboard as a graph, so the graph is a 9×9 grid. Is that graph hamiltonian? What about the graph that comprises the lines of a 7×7 board?

4.5. TREES

We shall illustrate the theoretical and practical importance of trees in this section. The definition we gave in Section 4.2 is only one of several possible ones, as we will now show.

Theorem 4.8. The following statements are equivalent for a graph G:

(a) G is a tree (i.e., connected and with no cycles).

(b) Every two vertices in G are joined by a unique path.

(c) G is connected and $p = q + 1$.

(d) G has no cycles and if any two nonadjacent vertices of G are joined by a new edge x, then $G + x$ has exactly one cycle.

Proof. **(a)** implies **(b)**. Since G is connected, every two vertices of G are joined by a path. If P_1 and P_2 are distinct paths joining u and v, let w be the first vertex on P_1 (traveling from u to v) so that w and its predecessors are on P_1 and P_2 but its successor is only on P_1. If w^* is the next vertex on P_1 which is also on P_2, then the sections on P_1 and P_2 between w and w^* together form a cycle. Since G is acyclic, this is impossible, so we cannot have both P_1 and P_2.

(b) implies **(c)**. If **(b)** holds then clearly G is connected. To show $p = q + 1$, we use induction on the number of vertices p. This is clearly true for $p = 1$ and 2. Assume it is true for graphs with at most $p - 1$ vertices. If G has p vertices, the removal of any edge x disconnects G, because of the uniqueness of paths, and in fact the new graph has exactly two components. By the induction hypothesis, each component has one more vertex than edge. If the two components have p_1 and p_2 vertices, and q_1 and q_2 edges, respectively, then $p_1 = q_1 + 1$, $p_2 = q_2 + 1$. Putting the edge x back in, we obtain $p = q + 1$, since $p_1 + p_2 = p$ and $q_1 + q_2 + 1 = q$.

(c) implies **(d)**. Suppose **(c)** holds and G contains a cycle of length n. There are n vertices and n edges on the cycle. For each vertex not on the cycle, there is a corresponding edge, namely the first edge on a shortest path between the vertex and the cycle. But this implies $q \geqslant p$, which is a contradiction. Thus G has no cycles, and is a tree, that is, **(a)** holds. Since **(a)** implies **(b)**, there is a unique path joining any pair of nonadjacent vertices u and v. Thus adding the edge $\{u, v\}$ to G creates a unique cycle in G.

(d) implies **(a)**. We have to prove that if **(d)** holds, then G is connected. If u and v lie in different components of G, then adding the edge $\{u, v\}$ cannot create a cycle in G. Thus G must be connected and is a tree.

Trees are important in their own right and also as subgraphs of more complex graphs. The following theorem is basic to many applications of trees. A subgraph of G is *spanning* if it contains every vertex of G.

Theorem 4.9. Every connected graph contains a spanning subgraph that is a tree.

Proof. If G has a cycle, remove any edge x of that cycle. This clearly cannot disconnect the graph. A finite number of repetitions of such edge-deletions will produce a subgraph that is connected and has no cycles, and is therefore the spanning subtree we are seeking.

Both Theorems 4.8 and 4.9 are useful in proving a classical result of Euler. A graph drawn in the plane without any edge-crossings is said to be *embedded* in the plane. An embedded planar graph generally divides the plane into several areas, called *faces*.

Euler's theorem relates the number of faces, edges, and vertices of a connected graph embedded in the plane. Because of strong historical tradition, we will depart from our usual notation for a moment and denote the number of vertices, edges, and faces of an embedded planar graph by V, E, and F, respectively.

Theorem 4.10. For any connected graph embedded in the plane,

$$V - E + F = 2. \tag{4.1}$$

Proof. Suppose G is embedded in the plane and T is a spanning tree of G. Then T has V vertices, $V - 1$ edges (by Theorem 4.8), and produces only one face in the plane, since it does not cut off any area from any other. Addition of any edge x produces a single cycle in $T + x$ and creates a new face in the plane. Another edge added to $T + x$ divides some region into two regions, thus increasing the number of faces by one. Continuing in this way, we get one more face each time we add an edge. After we have added all of the $E - (V - 1)$ edges that were not originally in the tree T, we have $1 + [E - (V - 1)]$ faces, so $F = 1 + [(E - (V - 1)]$. Simplifying, we get equation (4.1).

A second application of trees is to economical pipeline network design. Suppose that an oil field contains wells scattered over a large area. We want to design a pipeline network of minimum total length that connects all of the wells together, so that the oil can be collected at a single location. It is fairly clear that the network should be a tree, since cycles will serve no useful purpose. If we suppose further that we can put branching spots in the network only at wells (so that the wells form the vertices of our tree), there is a simple method to find the tree that utilizes the least pipe.

Theorem 4.11. Repetition of the following operation, until no edge satisfies the stated conditions, will result in a tree of smallest overall length spanning a given set of vertices. Choose a pair u, v of nonadjacent vertices, so that (a) the addition of the edge $\{u, v\}$ does not produce a cycle in the collection of edges chosen so far, and (b) the distance between u and v is minimal, among all pairs of vertices satisfying condition (a).

Proof. It is obvious that the procedure outlined will produce a spanning tree for the vertices. We must show that this tree has smallest total length. Call the tree we have constructed T.

Consider the set Ω of spanning trees with minimum overall length. Let T' be a tree in this set which has a maximum number of edges common with T. If $T = T'$, we are done. If not, let e be the first edge that was added to T which does not appear in T'. Adding e to T' produces a unique cycle C, by Theorem 4.8(d). Since T is a tree, there must be some edge f of T' in C that does not lie in T.

Let $T'' = T' - f + e$. Since f belongs to a cycle of $T' + e$, T'' must be connected. The number of edges and vertices in T'' is the same as in T. Hence from Theorem 4.8(c) T'' is a tree.

Let L be the set of edges of T chosen earlier than e. We show that the addition of f to L does not produce a cycle. If the addition of f to L produces a cycle, then since f and all the edges in L belong to T', this cycle is in T', which contradicts the fact that T' is a tree. If length $f <$ length e, then our algorithm would have chosen f instead of e. Hence length $f \geqslant$ length e. If length $f >$ length e, then T'' has a smaller length than T', which is a contradiction. Hence length $f =$ length e, that is $T'' \in \Omega$. This is a contradiction since T'' has more edges in common with T than does T'. Hence, in fact $T = T'$.

A procedure that solves a problem by a finite sequence of definite steps is called an *algorithm*. Although the method described in the theorem is not entirely definite (if two lengths are the same, which pair of vertices should be chosen?), most people would describe it as an algorithm. In fact, it is a "greedy" algorithm, because at each stage one chooses that which is best for the moment, adding the shortest possible edge. Note that length is only one possible criterion for choosing edges. More generally, one can attach a cost of any kind to each edge of a graph and the method of Theorem 4.11 will then produce a spanning tree of minimum total cost.

EXERCISES

1. Use Theorem 4.1 and the fact that a tree with p vertices has $p - 1$ edges to show that every tree has a vertex of degree 0 or 1.
2. Prove that every tree with at least two vertices has at least two end vertices (vertices of degree 1).
3. Show that a graph G is a tree if and only if it has no cycles and $p = q + 1$.
4. Draw all trees with 8 vertices. (There are 23 of them.)
5. Verify Theorem 4.10 for the graph of the dodecahedron, shown in Figure 4.11, Section 4.4.
6. Use Theorem 4.10 to show that the complete graph K_5 cannot be drawn in the plane without edges crossing. [*Hint*: If it was, how many faces would there be? How many edges are around each face?]
7. Find the least expensive way for transmitting information from x to all other nodes in the following graph. Edge labels indicate costs of transmission.

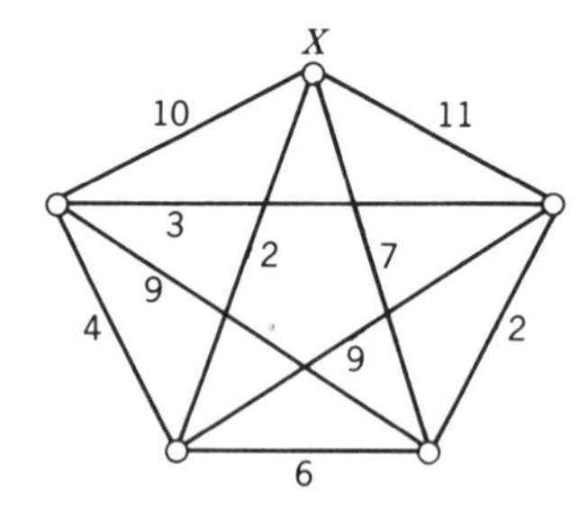

8. Find a shortest spanning tree to connect the oilwells shown below, with distances as indicated in the distance table.

Distances

From \ To	v_2	v_3	v_4	v_5	v_6	v_7	v_8
v_1	3	5	5	5	8	8	10
v_2		6	4	3	7	6	10
v_3			2	5	3	4	5
v_4				2	3	3	6
v_5					5	3	7
v_6						2	3
v_7							5

4.6. BINARY TREES

Trees turn up frequently in data handling with computers. In one very simple scheme for storing a collection of information, a *key* is attached at the beginning of each record (file) which helps to identify the record in question. At the end of each file we may append a pointer to indicate the location of the next file. Such an arrangement is known as a *linked list*, and clearly the underlying graphical structure is a path. Instead of one ponter, two can be appended, generally known as left and right, and then the graphical structure is a binary tree. Figure 4.14 shows a linked list and Figure 4.15a a binary tree. The numbers in the box containing the key indicate memory locations. Note that in each of these data structures the arrangement is in the natural order in some sense. Thus when the keys are names, the alphabetical order is natural. When the keys are numbers, we may naturally arrange them in ascending order of magnitude.

Suppose we want to set up a new set of files in the computer as a linked list, having keys JONES, SMITH, MOORE, BROWN, CARTER, and THOMAS, and they are to be input in the order indicated. Suppose the available memory locations are 1, 37, 19, 46, 13, and 81. We employ a START pointer to indicate, at all times, the beginning of the list. We put JONES in location 1 and set START to 1. We then put SMITH in the next location available, which is 37. Since SMITH comes after JONES alphabetically, we put SMITH after JONES in the diagram and set the pointer on JONES to 37. Since MOORE comes between JONES and SMITH in the alphabetical order we put him in memory location 19, and show him between JONES and SMITH in the diagram, and change the pointer on JONES to 19, and set the pointer on MOORE to 37. Figure 4.14 shows the linked list as it grows. When all the files have been input, the diagram is as in Figure 4.14(f).

The linked list is searched from the top. If we are looking for a record with key HART we use the START pointer to locate the beginning of the list, at memory location 46. We then compare HART to BROWN. Since

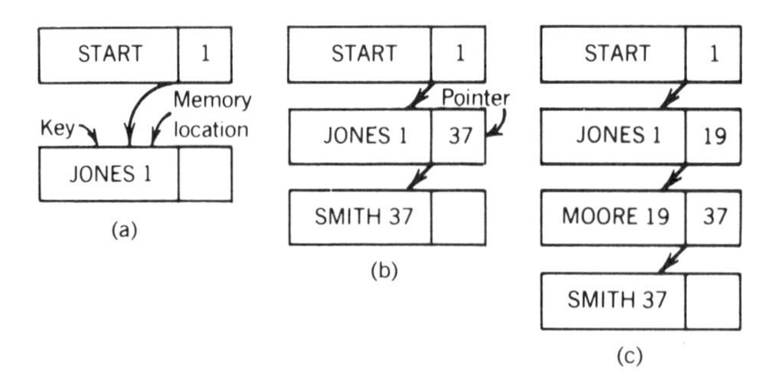

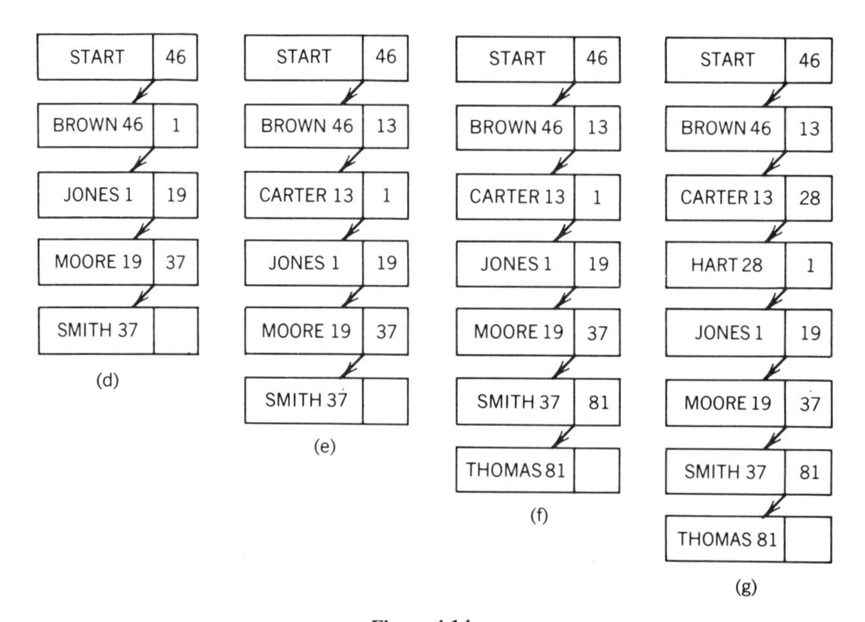

Figure 4.14

HART comes after BROWN, we go on to CARTER. Since HART comes after CARTER, we go on to the memory location 1 shown by the pointer on CARTER, thus coming to JONES. Since JONES should come after HART, we conclude that HART is not in the list. If we want to accommodate Hart in the list, we place HART in a vacant spot in the next memory location available, say location 28, change the pointer on CARTER to 28, and set the pointer on HART to 19. The diagram is now as in Figure 4.14(g). Note that when a new file is added to the list the location of any file already in the list does not change. Only some of the pointers have to be set on new numbers.

We will now explain how a binary tree grows. Suppose the keys to the records (files) are the same as in the previous example and they are to be input in the same order as before. We also assume that the same memory locations are available. Each file now has two pointers, the right and the left. We put JONES in location 1. We then put SMITH in the next location available. Since SMITH comes after JONES, we set the right pointer on Jones to 37. The next file is MOORE. Comparing it with JONES we find that MOORE comes after JONES. We therefore proceed to the location 37 shown by the right pointer on JONES, arriving at SMITH. Since MOORE comes before SMITH we put MOORE in the next location available, which is 19, and set the left pointer on SMITH to 19. The next file is BROWN. We compare BROWN with JONES. Since BROWN comes before JONES, we put BROWN in the next location available, which is 46, and set the left pointer on JONES (which is free, i.e., has not been set to any number yet) to 46. At this point our diagram looks as in Figure 4.15a. Both the right and left pointers on JONES have already been set, as also has the left pointer on SMITH. But the right pointer on SMITH and both the right and left pointers on MOORE are free. The next file is CARTER. We again start by comparing it with JONES. As CARTER is before JONES, we look at the left pointer on JONES and since it is on 46 we proceed to the location 46 arriving at BROWN. Now CARTER comes after BROWN, and the right pointer on BROWN is free. Hence we put CARTER in the next location available, which is 13, and set the right pointer on BROWN to 13. The next file is THOMAS. Since it comes after JONES we look at the right pointer on JONES and since it is on 37 we proceed to location 37 occupied by

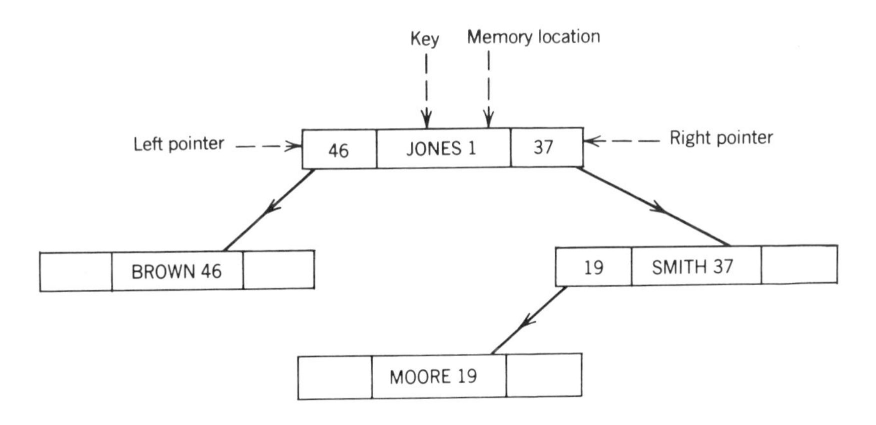

Figure 4.15a

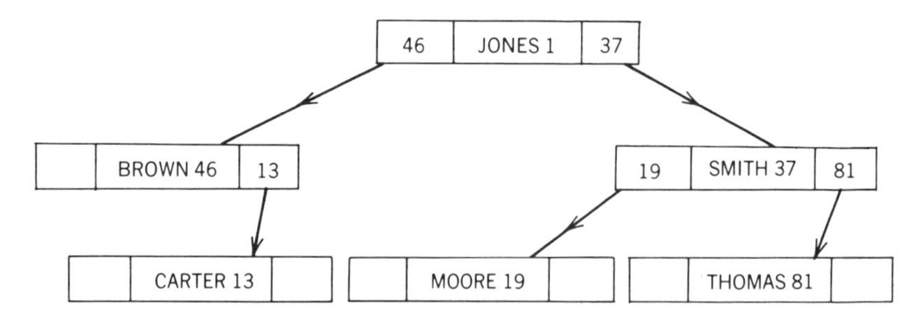

Figure 4.15b

SMITH. Since THOMAS comes after SMITH and the right pointer on SMITH is free we put THOMAS in the next location available, which is 81, and set the right pointer on SMITH to 81. Figure 4.15b is the diagram obtained when all six files have been input. No START pointer is required, since the initial vertex is never changed.

Note that the data structure obtained is that of a rooted directed binary tree. Each file together with the corresponding memory location can be thought of as a vertex. Each vertex is of incoming degree one, and the outgoing degree is two, one, or zero, respectively, as none, one, or both the pointers are free. As before, the binary tree is searched from the top (root). To search for HART we compare it with JONES. Since HART comes before JONES we look at the left pointer on JONES, which shows 46. We proceed to the location 46 occupied by BROWN. Since HART comes after BROWN, and the right pointer on BROWN shows 13, we proceed to the location 13 occupied by CARTER. Since both pointers on CARTER are free we conclude that HART is not in the list. To accommodate HART, we put it in the next location available, say 28, and since HART comes after CARTER we set the right pointer on CARTER to 28. We thus get the diagram of Figure 4.15c.

By the procedure explained before, linked lists and binary trees grow to handle any number of pieces of data. The advantage of the binary tree structure is that it generally requires fewer comparisons for a search, as we shall now see.

It is fairly obvious that a search of a linked list with n items will require, on the average, about $n/2$ comparisons to either locate a record or determine that it is not on file. A binary tree with a nice symmetrical shape, on the

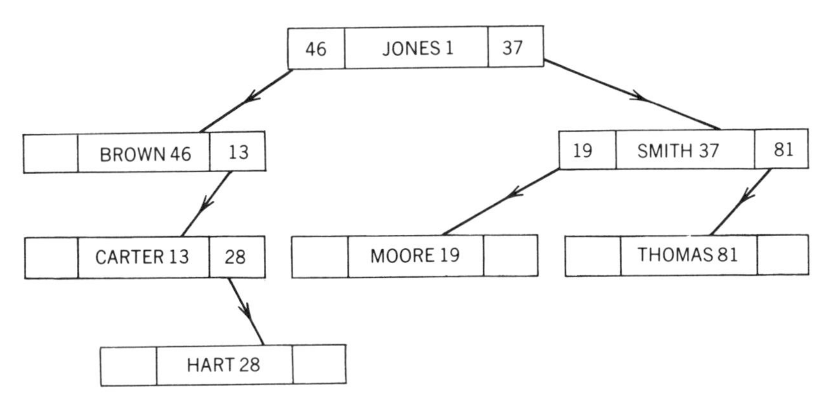

Figure 4.15c

other hand, requires only about $\log_2 n$ comparisons for an unsuccessful search, even less for a successful one. This is seen as follows.

Each vertex of the tree is at a certain level. The initial file (root) occupies the first level. The second level can accommodate up to two files corresponding to the left and right pointers of the initial file. The third level can accommodate up to four files, and so on. In general the nth level can accommodate up to a maximum of 2^{n-1} files, corresponding to the $2 \cdot 2^{n-2}$ pointers of the files in the preceding level (provided that the preceding level is fully occupied). Thus n levels can accommodate up to $2^n - 1$ files. If all the levels are fully occupied, then $2^n - 1$ files can be searched with n comparisons. This translates to $\log_2(2^n - 1)$ files being searched in $\log_2 n$ comparisons. Since $\log_2(2^n - 1) \approx n$ almost n files can be searched with $\log_2 n$ comparisons. A successful search will sometimes stop before the end of the process, and so requires even fewer comparisons, on the average. All of this may seem irrelevant, however, since we have no way to assure that our trees end up looking as nice as those in Figure 4.16. The process we have outlined for producing the trees by addition of records develops trees of various shapes, depending on the order in which data is added. In fact, however, it can be shown that if data are added in random order the expected number of comparisons needed to search the

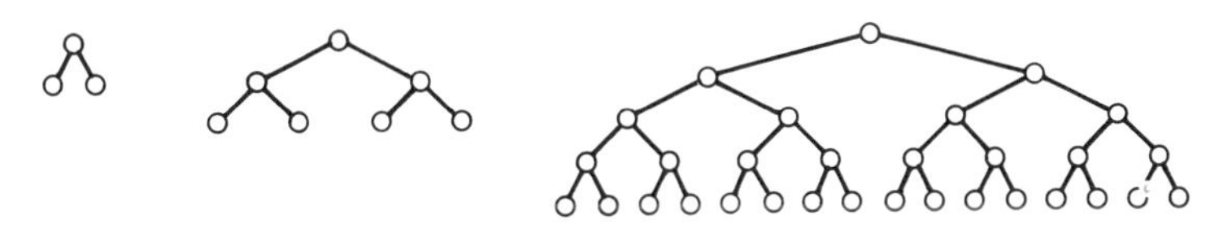

Figure 4.16

resulting binary tree is 2 ln n or approximately 1.386 $\log_2 n$. Thus random data in fact give trees that are fairly symmetrical. For large n, 2 ln n is far smaller than $n/2$ (for $n = 10$, 2 ln $n = 4.6$; for $n = 50$, 2 ln $n = 7.8$; for $n = 100$, 2 ln $n = 9.2$), so binary trees are far more efficient for data handling than are linked lists.

We conclude with another example of a binary tree. Suppose that the keys to the files are now the numbers 87, 41, 36, 86, 16, 70, 24, 91, 71, and 95 and they are to be input in this order. Suppose the memory locations are 1, 2, 3, 4, 5, 6, 7, 8, 9, and 10. In Figure 4.17 we have drawn the corresponding diagram, which illustrates better the binary tree nature of the data structure. The numbers in the circles, representing the vertices, are the keys. The memory location is written immediately below each key. Two directed lines proceed from each vertex. The line to the left of the vertical represents the left pointer and the line to the right of the vertical represents the right pointer. A pointer that is free is represented by a broken line ending with a circle containing X. For example, the left pointer of the file with key 91, in the memory location 8, is free is represented by a broken line ending with a circle containing X. For example, the left pointer of the file with key 91, in the memory location 8, is free. Other pointers are represented by solid lines, which are the edges of the binary tree and carry the numbers to which they are set. For example, the right pointer of the file 91, in the memory location 8, is on 10. This says that if in the process of search we arrive at the file 91, and the file to be searched has a number greater than 91, then we proceed to the memory location 10 shown by the right pointer on 91. This location contains the file with key 95. If we were searching for 93, then since 93 is less than 95 we will

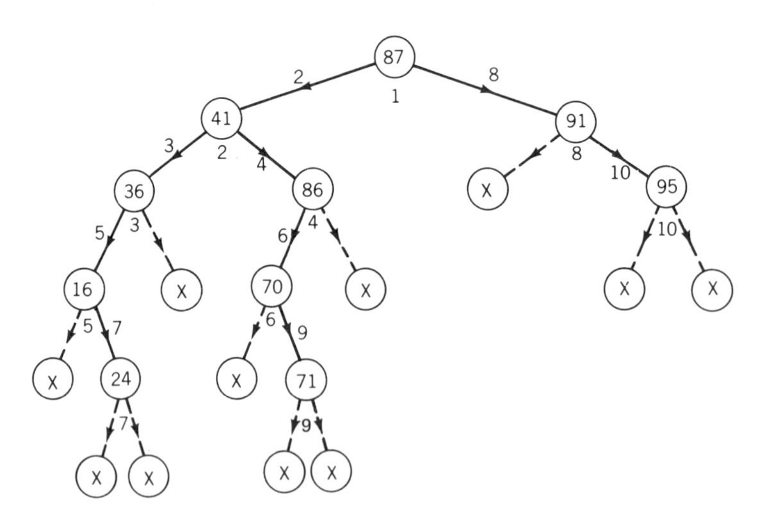

Figure 4.17

proceed to the location to which the left pointer of 95 is set. Since it is free, 93 is not present.

EXERCISES

1. Suppose that you are setting up a new set of files in the computer and have available memory locations 12, 27, 38, 42, 47, 81, 92, and 99. The files you want to store have keys JONES, THOMAS, FINK, HOWARD, LANG, BALL, and CRANE, and they are to be input in that order.
   (a) Draw a diagram showing how the data are stored as a linked list.
   (b) Draw a diagram showing how the data are stored as a binary tree.
2. What binary tree would result if the data in exercise 6 were input not in random order but alphabetically, beginning with BALL and ending with THOMAS?
3. There are 100 files for which the keys are the numbers 1 through 100. Ten of these were drawn in order at random and turned out to be 87, 41, 36, 86, 16, 70, 24, 91, 71, and 95. They were input in a computer as a binary tree, with the diagram in Figure 4.17. In the figure the unoccupied vertices shown by circles containing an X can be identified by the level and position, counting from the left.
   (a) If the next memory location is 11, and the file with the key 75 is drawn next, show that it must occupy the fourth free vertex at level 6.
   (b) For each of the free vertices write down the keys to the files, which if they happen to be drawn next must occupy that position.
   (c) Use (b) to find the average number of comparisons needed to locate a file drawn at random from the 100 files or to determine its absence. Compare this number with 2 ln 10.
   (d) After the first ten files have been drawn as indicated, six other files are drawn in order and turn out to be the files with keys 26, 45, 12, 48, 93, and 5. Extend the diagram of Figure 4.17 by accommodating these new files in the memory locations 11, 12, 13, 14, 15, and 16.
   (e) With respect to the extended data structure determine the average number of comparisons needed to locate a file or determine its absence. Compare this number with 2 ln 16.

4.7. REMARKS

The theory of graphs is a rich area of mathematics, with diverse applications. We have only been able to touch on a few topics which we hope illustrate the range (from obvious to very difficult) that graphical problems present. We have ignored whole areas, including enumeration of graphs and graph coloring.

Graph theory began in 1736, with Euler's work on what we now call eulerian trails. During the next 200 years there were few developments, although Kirchhoff used graphs to study electrical networks, Cayley counted trees in investigating hydrocarbons, and the Four-Color Conjecture for planar maps first appeared. The first textbook on the subject did not appear until 1936. Since then the theory has grown very rapidly, stimulated recently by applications to computers.

Readable texts on graph theory include Harary [1], Wilson [2], and Bollobás [3].

[1] F. Harary, *Graph Theory*, Addison-Wesley, Reading, 1969.

[2] R. J. Wilson, *Introduction to Graph Theory*, Academic Press, New York, 1972.

[3] B. Bollobás, *Graph Theory*, Springer-Verlag, New York, 1979.

CHAPTER 5

Finite Fields

5.1. INTRODUCTION

The familiar operations of addition (denoted by $+$) and multiplication (denoted by $\times$ or $\cdot$) on the system Z of integers clearly satisfy the following axioms:

A1. Existence of the sum.

Given any two elements a and b belonging to the system there is a unique element s belonging to the system, which is their sum:

$$s = a + b.$$

A2. The commutative law of addition.

$$a + b = b + a$$

A3. The associative law of addition.

$$(a + b) + c = a + (b + c).$$

A4. Existence of the difference or the law of subtraction.

Given any two elements a and b belonging to the system there is a unique element s belonging to the system, which is their sum:

$$a + x = b.$$

A5. Existence of the product.

Given any two elements a and b belonging to the system there is a unique element c belonging to the system, which is their product:

$$c = a \cdot b.$$

A6. The associative law of multiplication.

$$a \cdot (b \cdot c) = (a \cdot b) \cdot c$$

A7. The distributive law.

$$a \cdot (b + c) = a \cdot b + a \cdot c,$$

$$(b + c) \cdot a = b \cdot a + c \cdot a.$$

A8. The commutative law of multiplication.

$$a \cdot b = b \cdot a.$$

Again let M_n be the system of $n \times n$ matrices with real elements. Then we know that axioms A1–A7 are satisfied but A8 is not satisfied. We can think of other systems that satisfy only some of the axioms A1–A8, but may or may not satisfy the others. We thus adopt the following definitions:

A system of elements satisfying axioms A1–A4 is called a *module*. From the axioms A1–A4 we may deduce the following properties **(a)–(c)**. Every module must have these properties but they are not a part of our definition of a module since they can be deduced from the axioms defining a module.

(a) The element x in axiom A4 is unique.

(b) There exists a unique element 0 such that $a + 0 = a$, for any a belonging to the system.

(c) The element $-a$ is defined by $a + (-a) = 0$. We denote $b + (-a)$ by $b - a$. Then $a + x = b$, if and only if $x = b - a$. Also, $a - (-b) = a + b$.

A module is called a *ring* if it further satisfies the axioms A5–A7. Thus a system of elements is a ring if it satisfies axioms A1–A7. Note that a ring is necessarily a module but a module is not necessarily a ring. Thus the ring M_n of $n \times n$ matrices with real elements is a module but the set of n-vectors $(x_1, x_2, \ldots, x_n)$ with real elements is a module but not a ring, since the sum of $(x_1, x_2, \ldots, x_n)$ and $(y_1, y_2, \ldots, y_n)$ is defined by $(x_1 + y_1, x_2 + y_2, \ldots, x_n + y_n)$ and obeys axioms A1–A4 but a product satisfying A5–A7 is not defined.

From the axioms A1–A7 we may deduce the following properties **(d)–(f)**. Thus every ring satisfies properties **(a)–(f)**.

(d) $c \cdot 0 = 0 \cdot c = 0$.

(e) $c \cdot (-a) = (-c) \cdot a = -(c \cdot a)$.

(f) $b \cdot (c - a) = b \cdot c - b \cdot a$, $(c - a) \cdot b = c \cdot b - a \cdot b$.

It is customary to drop the dot from the product $a \cdot b$ and write it as ab when there is no ambiguity.

A ring is called a *commutative ring* if it further satisfies the axiom A8. Thus a commutative ring is necessarily a ring but the converse may or may not be true. Thus Z, the system of integers, is a commutative ring and hence is a ring. But M_n, the system of $n \times n$ matrices with real elements, is a ring but not a commutative ring.

The system of real numbers, besides satisfying the axioms A1–A8, also satisfies the following further axiom:

A9. Law of division.

If a and b are elements belonging to the system, and $a \neq 0$, then there exists an element y belonging to the system such that

$$ay = b.$$

Note that A9 is not necessarily true for a ring; for example, it does not hold for the commutative ring Z of integers.

A system for which the axioms A1–A9 are satisfied is called a *field*. Thus the system of real numbers is a field. Also the system of complex numbers is a field. On the basis of axioms A1–A9 we can prove the following properties of a field [of course, the properties **(a)**–**(f)** are also satisfied since every field is necessarily a ring]:

(g) The element y in A9 is uniquely determined.

(h) There exists a unique element 1 belonging to the system such that $a \cdot 1 = a$, for any a belonging to the system.

(i) $1 \neq 0$.

(j) If $ab = 0$, then at least one of a and b is equal to 0.

What interests us here is the existence of systems containing only a finite number of elements and yet satisfying all the axioms A1–A9. Such systems are called *Galois fields* after their discoverer Galois. We shall briefly sketch their properties here without attempting to give complete proofs.

EXERCISES

1. Prove that the properties **(a)**, **(b)**, and **(c)** hold for any module, being careful to assume nothing more than A1–A4.

2. Prove that the properties **(d)**, **(e)**, and **(f)** hold for any ring, being careful to assume nothing more than A1–A7.
3. Prove that the properties **(g)**, **(h)**, **(i)**, and **(j)** hold for any field, being careful to assume nothing more than A1–A9.
4. Prove that the set of all real numbers of the form $a + b\sqrt{2}$, where a and b are rational, form a field F. Find y such that $(3 + 5\sqrt{2})y = 2 - 3\sqrt{2}$, where $y \in F$.
5. Show that the set of all polynomials with real coefficients do not form a field. Do they form a ring? Is this ring commutative?
6. What is the zero element for the ring of 2×2 matrices with real entries? Give examples to show that the axioms A8 and A9 do not hold for this ring.
7. S is the set of complex numbers $a + bi$. S_1 is the subset of S for which both a and b are rational. S_2 is the subset of S for which a and b are integers (positive, negative, or zero). Is the axiom A9 for fields satisfied in S_1? Is it satisfied in S_2? For $a = 3 + 8i$ and $b = 2 + 3i$ belonging to S_1, find y belonging to S_1, such that $ay = b$.

5.2. THE GALOIS FIELD, GF$_p$

Let g be a positive integer. Any two integers a and b (positive, negative, or zero) are said to be *congruent modulo* g if $a - b$ is divisible by g, and this fact is denoted by

$$a \equiv b \pmod{g}.$$

If a and b are not congruent modulo g, then they are said to be *incongruent modulo* g, and this is denoted by

$$a \not\equiv b \pmod{g}.$$

For example, $15 \equiv 1 \pmod{7}$, because $15 - 1 = 14$ is divisible by 7. Also $-3 \equiv 11 \pmod{7}$ because $-3 - 11 = -14$ is divisible by 7. But $13 \not\equiv 2 \pmod{7}$ since 11 is not divisible by 7.

Lemma 5.1. If $a \equiv b \pmod{g}$ and $b \equiv c \pmod{g}$, then $a \equiv c \pmod{g}$.

Proof.

$$a - b = q_1 g, \quad b - c = q_2 g,$$

where q_1 and q_2 are integers. Hence

$$a - c = (q_1 + q_2)g,$$
$$a \equiv c(\text{mod } g).$$

Also a is congruent to itself (mod g), and if $a \equiv b$ (mod g) then $b \equiv a$ (mod g). Hence we can divide the set of integers into classes such that any two integers belonging to the same class are congruent to each other.

The class to which the integer a belongs may be denoted by (a). Thus if a and b belong to the same class then $(a) = (b)$. For example, if $g = 7$, then $(15) = (1), (-3) = (11)$. We can write

$$a = a_1 + qg, \qquad \text{where } 0 \leqslant a_1 < g,$$

and q is an integer. Then a_1 may be called the *standard representative* of the class (a). Clearly $(a) = (a_1)$. For example, $15 = 1 + 2 \cdot 7$, $-3 = 4 - 1 \cdot 7$, $-17 = 4 - 3 \cdot 7$. Hence, 1 is the standard representative of (15), and $(15) = (1)$. Similarly $(-3) = (4), (-17) = (4)$. Since there are exactly g integers satisfying $0 \leqslant a_1 < g$ there are exactly g different classes (mod g). They are called *residue classes* (mod g). These g different classes are

$$(0), (1), (2), \ldots, (g - 1).$$

Consider the system of residue classes (mod g). We define the operations of addition and multiplication in this system by the rules

$$(a) + (b) = (a + b),$$
$$(a) \cdot (b) = (ab).$$

For example, if $g = 7$, then

$$(5) + (6) = (11) = (4),$$
$$(5) \cdot (6) = (30) = (2).$$

It is readily seen that the system of residue classes (mod g) is a ring. We thus get an example of a finite ring. We shall now prove the following theorem.

Theorem 5.1. If p is a prime then the system of residue classes (mod p) is a field.

Proof. Since the system is a ring we have only to prove that the axiom

A9, that is, the law of division holds. Let $(a) \neq 0$, and (b) be any two residue classes (mod p). We have then to find a class (y) such that $(a)(y) = (b)$. Let

$$\begin{aligned}(a)\cdot(0) &= (b_0)\\ (a)\cdot(1) &= (b_1)\\ (a)\cdot(2) &= (b_2)\\ \vdots \quad & \quad \vdots \\ (a)(p-1) &= (b_{p-1}).\end{aligned}$$

We may take the classes (b_i) in the standard form, that is, $0 \leqslant b_i < p$. We claim that the classes (b_i), where $i = 0, 1, 2, \ldots, p-1$, are all different. If possible suppose $(b_i) = (b_j)$, where $0 \leqslant i < j \leqslant p - 1$. Then $(a)\cdot[(j) - (i)] = (0)$, and so (p) divides $(a)\cdot[(j) - (i)]$. Since p is prime and does not divide a, this implies p divides $j - i$, which is impossible. This proves our claim. Therefore, exactly one of the classes $(b_0), (b_1), \ldots, (b_{p-1})$ is equal to (b), say (b_i). Then the required class $(y) = (i)$.

The field of residue classes (mod p) is denoted by GF_p, where the letters GF stand for Galois field.

When the modulus is understood we can drop the parenthesis and write the class (x) as x. Thus if $p = 7$, we can write

$$3\cdot 4 = 5, \qquad 2 + 6 = 1,$$

and so on. All the standard operations that are possible in a field can be carried out in GF_p. In particular we can solve linear equations.

Example 1. For the field GF_7 solve the linear equations

$$\begin{aligned}3x + 2y &= 5,\\ 5x + 4y &= 4.\end{aligned}$$

Multiplying the first equation by 5 and the second equation by 3 we get

$$\begin{aligned}x + 3y &= 4,\\ x + 5y &= 5.\end{aligned}$$

Hence by subtraction $2y = 1$ or $y = 4$ and $x = 6$.

EXERCISES

1. Write out the complete summation and multiplication tables for the Galois field GF_5.
2. Show that if g is *not* prime then the system of residue classes (mod g) do *not* form a field.
3. In the system of residue classes (mod g), which elements have multiplicative inverses? (Two elements are said to be multiplicative inverses if their product is 1.)
4. In GF_7 compute the standard representatives of:
   (a) -3
   (b) $3 \cdot 6$
   (c) $5/6$
   (d) 4^{-1}
   (e) $\dfrac{3^2 - 5 \cdot 4}{4^3 + 5^2}$
   (f) $\begin{vmatrix} 3 & 2 & 1 \\ 4 & 1 & 5 \\ 2 & 3 & 6 \end{vmatrix}$
5. In GF_7 solve the equations

$$\begin{aligned} x + 2y + 3z &= 2, \\ 2x + 4y + 5z &= 6, \\ 3x + y + 6z &= 4. \end{aligned}$$

5.3. THE COMMUTATIVE RING $GF_p[x]$

We shall consider polynomials

$$f(x) = a_0 + a_1x + a_2x^2 + \cdots, \tag{5.1}$$

where the coefficients belong to the field GF_p. The highest power of x having a nonzero coefficient is called the *degree* of the polynomial. The addition and multiplication of polynomials is defined in the usual way. Thus if $f(x)$ is given by (5.1) and

$$g(x) = b_0 + b_1x + b_2x^2 + \cdots,$$

then

$$f(x) + g(x) = (a_0 + b_0) + (a_1 + b_1)x + (a_2 + b_2)x^2 + \cdots,$$
$$f(x)g(x) = a_0b_0 + (a_0b_1 + a_1b_0)x + (a_0b_2 + a_1b_1 + a_2b_0)x^2 + \cdots.$$

Under these operations the polynomials form a commutative ring denoted by $GF_p[x]$.

Example 2. Let $p = 5$, and let

$$f(x) = 2 + 3x + x^3,$$
$$g(x) = 1 + 4x + x^2 + 4x^3.$$

Then each of $f(x)$ and $g(x)$ has degree 3, and

$$f(x) + g(x) = 3 + 2x + x^2,$$
$$f(x)g(x) = 2 + x + 4x^2 + 2x^3 + x^4 + x^5 + 4x^6.$$

The elements of GF_p may be regarded as polynomials of degree zero.

If $f(x)$ and $g(x)$ are two polynomials of $GF_p[x]$ with degree $f(x) \geqslant$ degree $g(x)$ then we divide $f(x)$ by $g(x)$ and express $f(x)$ in the form

$$f(x) = q(x)g(x) + r(x),$$

where deg $r(x) <$ deg $g(x)$. $q(x)$ is the quotient and $r(x)$ is the remainder obtained on dividing $f(x)$ by $g(x)$. If $r(x) = 0$, then $f(x) = q(x)g(x)$ and we say that $f(x)$ is divisible by $g(x)$.

Example 3. Let $p = 5$,

$$f(x) = x^4 + 3x^3 + 2x^2 + x + 3,$$
$$g(x) = x^2 + 4x + 2.$$

We illustrate the process of division.

$$\begin{array}{r l}
 & x^2 + 4x + 4 = q(x) \\
g(x) = x^2 + 4x + 2 & \overline{\big)\, x^4 + 3x^3 + 2x^2 + x + 3} = f(x) \\
 & \underline{x^4 + 4x^3 + 2x^2} \\
 & \qquad 4x^3 \qquad\qquad + x
\end{array}$$

$$\begin{array}{r} 4x^3 + x^2 + 3x \\ \hline 4x^2 + 3x + 3 \\ 4x^2 + x + 3 \\ \hline 2x \end{array} = r(x).$$

Thus

$$f(x) = q(x)g(x) + r(x),$$

where

$$q(x) = x^2 + 4x + 4, \qquad r(x) = 2x.$$

A polynomial $f(x)$ of $GF_p[x]$ is said to be *irreducible* over GF_p if it is impossible to find polynomials $\phi_1(x)$ and $\phi_2(x)$ of $GF_p[x]$ of degrees m and n with $m \geqslant 1$, $n \geqslant 1$ such that

$$f(x) = \phi_1(x)\phi_2(x).$$

If, however, we can find $\phi_1(x)$ and $\phi_2(x)$ satisfying the above conditions, then $f(x)$ is said to be *reducible* over GF_p. In this case $f(x)$ is divisible by both $\phi_1(x)$ and $\phi_2(x)$.

Example 4. The polynomial $x^2 + 1$ of $GF_5[x]$ is reducible over GF_5 since we can write

$$x^2 + 1 = (x + 2)(x + 3).$$

On the other hand, it is easy to verify that the polynomial $x^2 + 2$ of $GF_5[x]$ is irreducible over GF_5.

EXERCISES

1. Find, in $GF_7[x]$,

(a) $(x^2 + 6x + 1) - (x^3 + x + 5)$,

(b) $(x^3 + x^2 + 4) + (x^6 + 6x^2 + x)$,

(c) $(x^3 + x^2 + 4) \cdot (x^3 + 2x + 5)$.

2. Decide, for each of the following polynomials, whether or not it is reducible over (i) GF_2 and (ii) GF_7.

(a) $x^2 + x + 1$,

(b) $x^3 + x + 1$,
(c) $x^2 + 1$,
(d) $x^3 + x^2 + x + 1$.

In the case of reducibility determine the factors.

3. Determine all possible irreducible monic polynomials of degree 2 over
   (a) GF_3,
   (b) GF_5.
4. In the ring $GF_7[x]$, divide the polynomial $f(x) = x^4 + 3x^3 + 5x^2 + 6$ by $\phi(x) = x^2 + 2x + 5$. Hence express $f(x)$ in the form $f(x) = q(x)\phi(x) + r(x)$.

5.4. THE GALOIS FIELD GF_{p^n}

Let $\phi(x)$ be a polynomial of $GF_p[x]$. The polynomials $f_1(x)$ and $f_2(x)$ are said to be *congruent modulo* $\phi(x)$ if $f_1(x) - f_2(x)$ is divisible by $\phi(x)$. In this case we write

$$f_1(x) \equiv f_2(x) \quad [\text{mod } \phi(x)].$$

For example, if $p = 5$,

$$f_1(x) = x^3 + 2x^2 + x,$$
$$f_2(x) = x^3 + x^2 + x + 4,$$

then

$$f_1(x) \equiv f_2(x) \quad [\text{mod } (x + 2)],$$

since $f_1(x) - f_2(x) = x^2 + 1$ is divisible by $x + 2$.

For a given modulus polynomial $\phi(x)$, the class of all polynomials congruent to $f(x)$ may be denoted by $[f(x)]$. Addition and multiplication of these classes may then be defined by

$$[f_1(x)] + [f_2(x)] = [f_1(x) + f_2(x)],$$
$$[f_1(x)][f_2(x)] = [f_1(x)f_2(x)].$$

When the modulus polynomial $\phi(x)$ is irreducible over GF_p, it is known that the classes defined above form a field. Let the degree of $\phi(x)$ be n. Let $[f(x)]$ be any class. Using division we can write

$$f(x) = \phi(x)q(x) + f_1(x),$$

where $\deg f_1(x) < \deg \phi(x)$. Thus $\deg f_1(x) \leqslant n - 1$. Then $f_1(x)$ may be said to be the *standard representative* of $[f(x)]$. We may write

$$f_1(x) = a_0 + a_1 x + a_2 x^2 + \cdots + a_{n-1} x^{n-1}.$$

Since a_i belongs to GF_p, a_i can take p different values. Hence the number of standard representatives and, therefore, the number of residue classes modulo $\phi(x)$ is p^n. Thus the field defined by GF_p and $\varphi(x)$ contains exactly p^n different elements. We say that p^n is the *order* of the field.

Example 5. Consider the commutative ring $GF_3[x]$. It is easily checked that $x^2 + 1$ is irreducible over GF_3. Thus the field of residue classes [mod $(x^2 + 1)$] contains nine elements:

$[0], [1], [2], [x], [x + 1], [x + 2], [2x], [2x + 1], [2x + 2]$.

(i) $[x] + [x + 1] = [2x + 1]$.
(ii) $[2x + 1] + [2x + 2] = [x]$.
(iii) $[x + 1][x + 2] = [x^2 + 2] = [1]$.
(iv) $[x + 1][2x + 2] = [2x^2 + x + 2] = [x]$.

It is known that any field with a finite number of elements contains p^n elements, where p is a positive prime and n is a positive integer. Two finite fields with the same number of elements are isomorphic, that is, structurally identical. We can make a (1, 1) correspondence between their elements in such a way that the sum of two elements corresponds to the sum of corresponding elements, and the same holds for the product of the two elements. The finite field or Galois field with p^n elements may be symbolized by GF_{p^n}.

Example 6. To illustrate that there is essentially one field with p^n elements let us take $p = 3$, $n = 2$. Instead of the irreducible polynomial $x^2 + 1$, which was chosen as the modulus in Example 5, we can choose another irreducible polynomial, say $f(x) = x^2 + x + 2$. We again get nine classes

$$\{0\}, \{1\}, \{2\}, \{x\}, \{x + 1\}, \{x + 2\}, \{2x\}, \{2x + 1\}, \{2x + 2\}.$$

The correspondence between these classes and the classes of Example 5 is given by

$$[0] \to \{0\},$$
$$[1] \to \{1\},$$
$$[2] \to \{2\},$$

$$\begin{aligned}
[x] &\to \{x + 2\},\\
[x + 1] &\to \{x\},\\
[x + 2] &\to \{x + 1\},\\
[2x] &\to \{2x + 1\},\\
[2x + 1] &\to \{2x + 2\},\\
[2x + 2] &\to \{2x\}.
\end{aligned}$$

Now

(i) $\{x + 2\} + \{x\} = \{2x + 2\}$,
(ii) $\{2x + 2\} + \{2x\} = \{x + 2\}$,
(iii) $\{x\}\{x + 1\} = \{x^2 + x\} = \{1\}$,
(iv) $\{x\}\{2x\} = \{2x^2\} = \{x + 2\}$.

Compare the operations **(i)**, **(ii)**, **(iii)**, and **(iv)** of Example 5 with the operations **(i)**, **(ii)**, **(iii)**, and **(iv)** of Example 6. The elements on the left correspond to the elements on the left and the elements on the right correspond to the elements on the right. A similar result holds for both addition and multiplication of any pair of elements.

EXERCISES

1. Consider the polynomial $x^3 + x + 1$ in $GF_2[x]$.
   (a) Show this polynomial is irreducible.
   (b) In the field $GF_2[x]$ mod $(x^3 + x + 1)$ calculate:
      (i) $(x^2 + 1) + (x^2 + x + 1)$,
      (ii) $x \cdot (x^2 + x + 1)$,
      (iii) $(x^2 + 1) \cdot (x^2 + x)$,
      (iv) $(x + 1)^{-1}$,
      (v) $(x + 1)/(x^2 + x + 1)$.

2. Find *two* fields of order 8 by using GF_2 with the irreducible polynomials $x^3 + x^2 + 1$ and $x^3 + x + 1$. Show they are isomorphic by exhibiting a correspondence between their classes which is one-to-one and preserves sums and products.

3. Show that $x^2 + 3$ is irreducible over GF_5. Hence construct a field of 25 elements $ax + b$, where a and b belong to GF_5. In this field show that if

$$\frac{ax + b}{cx + d} = mx + n,$$

then m and n are uniquely determined by the equations

$$dm + cn = a,$$
$$2cm + dn = b.$$

In this field find the value of

$$\frac{(2x + 3)(4x + 1) + (3x + 2)}{(x + 4) - (3x + 2)}.$$

5.5. PRIMITIVE ELEMENTS OF GF_{p^n}

It is known that every nonzero element θ of GF_{p^n} (i.e., the finite field with p^n elements) satisfies the equation

$$\theta^{p^n - 1} = 1.$$

A nonzero element θ is said to be a *primitive element* of the field if all the powers of θ less than $p^n - 1$ are different. Thus, if θ is a primitive element, then

$$\theta^0 (= 1), \theta, \theta^2, \theta^3, \ldots, \theta^{p^n - 2}$$

are all different and these are all the elements. In particular $\theta^i \neq 1$, for $0 < i < p^n - 1$. The class $[x]$ may or may not be a primitive element according to the choice of the modulus.

Example 7. Consider the field GF_{3^2} obtained by taking the polynomial $\phi(x) = x^2 + 1$ as the modulus polynomial as in Example 5. Note that $[x]^i = [x^i]$. Now

$$[x]^2 = [x^2] = [2],$$
$$\therefore [x]^4 = [2]^2 = [4] = [1].$$

But $p^n - 1 = 8$. Hence $[x]$ is not a primitive element.

On the other hand, let our modulus polynomial be $\psi(x) = x^2 + x + 2$ as in Example 6. Then

$$\begin{aligned}
[x^o] &= 1, \\
[x^1] &= [x], \\
[x^2] &= [2x + 1], \\
[x^3] &= [2x^2 + x] = [2x + 2], \\
[x^4] &= [2x^2 + 2x] = [2], \\
[x^5] &= [2x], \\
[x^6] &= [2x^2] = [x + 2], \\
[x^7] &= [x^2 + 2x] = [x + 1].
\end{aligned}$$

Hence, the powers of $[x]$ less than 8 are all different. Hence, in this case $[x]$ is a primitive element. Note

$$[x^8] = [x^2 + x] = [1],$$

as it should be.

It is known that there always exists a modulus polynomial such that when we use it to construct the field GF_{p^n}, then the class $[x]$ is a primitive element. Such a polynomial is called a *minimum function*. We give below a table of minimum functions for a few fields:

Field	Minimum Function
GF_{2^2}	$x^2 + x + 1$
GF_{2^3}	$x^3 + x^2 + 1$
GF_{3^2}	$x^2 + x + 2$.

In what follows we shall always choose an appropriate minimum function as the modulus polynomial in constructing a finite field GF_{p^n}. The prime p is called the *characteristic* of the field. Once the characteristic and the minimum functions are given, we shall drop the parenthesis and write $[f(x)]$ simply as $f(x)$. In particular, for $[x]$ we may write x. Then every element of GF_{p^n} can be written in the form of a polynomial

$$a_0 + a_1 x + \cdots + a_{n-1} x^{n-1}$$

of degree $n - 1$ or less. Also, every nonzero element can be written as a power of x. This greatly facilitates calculations as will be illustrated in the next section.

EXERCISES

1. For the field GF_{3^2} show that $x^2 + x + 2$ is a minimum function.
2. For the field GF_{3^2} show that $x^2 + 1$ is not a minimum function, and find an element of GF_{3^2} which is a primitive element (mod $x^2 + 1$).
3. Verify that $x^4 + x + 1$ is a minimum function for GF_{2^4} and use it to exhibit the elements of GF_{2^4} as powers of the primitive element x and as polynomials of degree 3 or less in x with coefficients from GF_2.
4. Show that $f(x) = x^3 + 2x + 1$ is a minimum function for GF_{2^3}. Show that $\phi(x) = x^3 + x^2 + x + 2$ is irreducible over GF_3. Is $\phi(x)$ a minimum function? Obtain the isomorphism between the field of residue classes mod $f(x)$ and the residue classes mod $\phi(x)$ of the polynomials of the ring $GF_3[x]$.
5. Show that $f(x) = x^2 + 2x + 3$ is a minimum function for GF_{5^2}. Express the elements of GF_{5^2} as powers of the primitive element x and as linear polynomials $ax + b$, where a and b belong to GF_5.

5.6. OPERATIONS IN GF_{p^n}

In this section we display the arithmetic of four different finite fields.

(a) The Field GF_{3^2}

Choosing the minimum function $f(x)$ as $x^2 + x + 2$, we can write the nine elements as

$$\begin{array}{lll}
\alpha_0 = 0 = 0, & \alpha_3 = x^2 = 2x + 1, & \alpha_6 = x^5 = 2x, \\
\alpha_1 = x^0 = 1, & \alpha_4 = x^3 = 2x + 2, & \alpha_7 = x^6 = x + 2, \\
\alpha_2 = x^1 = x, & \alpha_5 = x^4 = 2, & \alpha_8 = x^7 = x + 1, \\
 & x^8 = 1. &
\end{array}$$

Thus each element has two forms.

The expression as a power of x is called the *multiplicative form*, and the expression as a polynomial of degree less than the minimum function is called the *additive form*. In carrying on the operations of multiplication and division we use the multiplicative form and remember that

$$x^{p^n - 1} = 1.$$

In carrying on the operations of addition and subtraction we use the additive form and remember that the coefficients are added and subtracted (mod p).

$$\begin{aligned}
\alpha_5\alpha_8 &= x^4x^7 = x^{11} = x^3 = \alpha_4, \\
\alpha_2/\alpha_6 &= x/x^5 = x^{-4} = x^4 = \alpha_5, \\
\alpha_7/\alpha_4 &= x^6/x^3 = x^3 = \alpha_4, \\
\alpha_3/\alpha_8 &= x^2/x^7 = x^{-5} = x^3 = \alpha_4, \\
\alpha_7 + \alpha_6 &= (x + 2) + 2x = 2 = \alpha_5, \\
\alpha_5 - \alpha_3 &= 2 - (2x + 1) = x + 1 = \alpha_8.
\end{aligned}$$

(b) The Field GF_{2^2}

We shall use the minimum function $x^2 + x + 1$. Then the four elements are

$$\begin{aligned}
\alpha_0 &= 0, \\
\alpha_1 &= 1, \\
\alpha_2 &= x, \\
\alpha_3 &= x^2 = x + 1, \\
x^3 &= 1.
\end{aligned}$$

Note that in a field of characteristic 2, $+1 = -1$. Hence, addition is the same as subtraction, that is $a + b = a - b$. Also $2a = 0$.

The field GF_{2^2} is small enough to write down the full addition and multiplication tables:

Addition table

	α_0	α_1	α_2	α_3
α_0	α_0	α_1	α_2	α_3
α_1	α_1	α_0	α_3	α_2
α_2	α_2	α_3	α_0	α_1
α_3	α_3	α_2	α_1	α_0

Multiplication table

	α_1	α_2	α_3
α_1	α_1	α_2	α_3
α_2	α_2	α_3	α_1
α_3	α_3	α_1	α_2

Since α_0 is zero the product of any element with α_0 is also α_0.

Example 8. Solve the equations

$$\alpha_2 x + \alpha_3 y = \alpha_1,$$
$$\alpha_3 x + \alpha_2 y = \alpha_3.$$

Multiplying the equations by α_3 and α_2, respectively, we have

$$\alpha_3\alpha_2 x + \alpha_2 y = \alpha_3,$$
$$\alpha_2\alpha_3 x + \alpha_3 y = \alpha_1,$$

subtracting

$$\alpha_1 y = \alpha_2,$$

hence

$$y = \alpha_2.$$

Substituting this in the first equation we find

$$\alpha_2 x = \alpha_1 - \alpha_2\alpha_3 = \alpha_1 + \alpha_1 = 0.$$

Hence, $x = 0 = \alpha_0$. Thus

$$(x, y) = (\alpha_0, \alpha_2).$$

We may check the correctness of the solution by actual substitution.

(c) The Field GF_{2^3}

The modulus polynomial can now be taken as $\psi(x) = x^3 + x^2 + 1$. The eight elements can be expressed as

$$\begin{aligned}
\alpha_0 &= 0, & \alpha_4 &= x^3 = x^2 + 1, \\
\alpha_1 &= 1, & \alpha_5 &= x^4 = x^3 + x = x^2 + x + 1, \\
\alpha_2 &= x, & \alpha_6 &= x^5 = x^3 + x^2 + x = x + 1, \\
\alpha_3 &= x^2, & \alpha_7 &= x^6 = x^2 + x, \\
& & x^7 &= 1.
\end{aligned}$$

Note that this again is a field of characteristic 2.

Example 9. Find the value of $(\alpha_3\alpha_4 + \alpha_7)/\alpha_5\alpha_2$.

$$\begin{aligned}
\frac{\alpha_3\alpha_4 + \alpha_7}{\alpha_5\alpha_2} &= \frac{\alpha_6 + \alpha_7}{\alpha_6}, && \text{since } \alpha_3\alpha_4 = x^2x^3 = x^5 = \alpha_6, \\
& && \alpha_5\alpha_2 = x^4x = x^5 = \alpha_6; \\
&= \frac{\alpha_4}{\alpha_6}, && \text{since } \alpha_6 + \alpha_7 = (x+1) + (x^2 + x) \\
& && = x^2 + 1 = \alpha_4; \\
&= \alpha_6, && \text{since } \alpha_4/\alpha_6 = x^3/x^5 = x^{-2} = x^5 = \alpha_6.
\end{aligned}$$

(d) The Field GF_{19}

In the field of residue classes (mod p), that is, the field GF_p, the operations of addition, subtraction, and multiplication are easily carried out but division is cumbersome as we have to do a lot of testing. Thus, in GF_{19} to find y such that

$$12y = 16$$

we will have to substitute the different values of y to see which one gives the correct result. This can be avoided by constructing a table of the powers of a primitive root. We know from theory of numbers that for any prime p there always exists a positive integer x, such that the powers of x less than $p - 1$ taken (mod p) are all different, and $x^{p-1} \equiv 1 \pmod{p}$. x is called a *primitive root* of p.

In GF_p we express the powers of a primitive root x in the standard form and use this table as a help in division. For example, for $p = 19$ we know that 2 is a primitive root. We write

$$\begin{array}{llllll}
2^0 = 1, & 2^1 = 2, & 2^2 = 4, & 2^3 = 8, & 2^4 = 16, & 2^5 = 13, \\
2^6 = 7, & 2^7 = 14, & 2^8 = 9, & 2^9 = 18, & 2^{10} = 17, & 2^{11} = 15, \\
2^{12} = 11, & 2^{13} = 3, & 2^{14} = 16, & 2^{15} = 12, & 2^{16} = 5, & 2^{17} = 10,
\end{array}$$
$$2^{18} = 1.$$

To find y such that $12y = 16$, we now have

$$y = \frac{16}{12} = \frac{2^4}{2^{15}} = 2^{-11} = 2^7 = 14.$$

We give below a table of primitive roots for some values of p.

p	Primitive Root
3	2
5	2
7	3
11	2
13	2
17	3
19	2
23	5

EXERCISES

1. If the elements of GF_{2^3} are expressed as in Section 5.6(c), solve the equations

$$\alpha_4 x + \alpha_3 y = \alpha_7,$$
$$\alpha_5 x + \alpha_2 y = \alpha_3.$$

2. If the elements of GF_{3^2} are expressed as in Section 5.6(a), find all solutions (x, y) to the linear equation $\alpha_4 x + \alpha_8 y = \alpha_1$.

3. If the elements of GF_{2^2} are expressed as in Section 5.6(b), calculate the value of the determinant

$$\begin{vmatrix} \alpha_1 & \alpha_2 & \alpha_3 \\ \alpha_3 & \alpha_1 & \alpha_2 \\ \alpha_2 & \alpha_3 & \alpha_1 \end{vmatrix}$$

4. Verify that 5 is a primitive root of 23 by expressing powers of 5 (mod 23). If i represents the residue class (i), (mod 23) find the value of

$$13\left(\frac{6}{19} + \frac{15}{11}\right) - \frac{3}{22}.$$

5. If the elements of GF_{2^4} are expressed as in exercise 3 of Section 5.5, find the value of

$$\frac{[(x^3 + x^7)\,x^{13} + x^{11}]}{x^4 + x^5}.$$

5.7. REMARKS

We have seen that the arithmetic properties that we usually associate with the integers can be observed in certain finite sets if operations are defined carefully. We have just touched on the richness of abstract algebra, a great branch of mathematics that studies groups, rings, and other objects as well as fields. The simple results of this chapter will be applied repeatedly in the next three chapters to construct various combinatorial structures.

Finite fields were first used by Galois in an 1830 paper on the solution of certain equations. Unfortunately, Galois was killed in a duel in 1832, at the tender age of 20, and his work was little understood for many years. Abstract algebra did not really develop until the end of the 19th century, when work by Cauchy, Hamilton, Cayley, Moore, Dickson, and others initiatied the burst of activity that continues to this day.

If you want to learn more about finite fields, you can consult a textbook on abstract algebra. One such book that gives a particularly nice treatment of finite fields is by Dean [1], although Herstein's text [2] may be easier to find. Books on coding theory, such as Berlekamp [3] and Peterson and Weldon [4], also often contain useful chapters on Galois fields.

[1] R. A. Dean, *Elements of Abstract Algebra*, Wiley, New York, 1966.

[2] I. N. Herstein, *Topics in Algebra*, Xerox, Lexington, 1975.

[3] E. R. Berlekamp, *Algebraic Coding Theory*, McGraw-Hill, New York, 1968.

[4] W. W. Peterson and E. J. Weldon, Jr., *Error-Correcting Codes*, MIT Press, Cambridge, 1972.

CHAPTER 6

Finite Plane Geometries

Finite fields that have been studied in the last chapter may be used to construct geometrical systems quite similar to those based on real and complex fields. In this chapter we shall consider finite planes that, in many respects, resemble ordinary Euclidean or projective geometry.

6.1. THE GEOMETRY EG(2, p^n)

The finite *Euclidean geometry* of two dimensions over the field GF_{p^n} is denoted by EG(2, p^n) and defined as follows.

There are two kinds of elements: "points" and "lines." A *point* is a pair (x, y), where $x \in GF_{p^n}$, $y \in GF_{p^n}$. The numbers x and y are the *coordinates of the point.*

A *line* is the set of points whose coordinates satisfy a linear equation

$$ax + by + c = 0, \tag{6.1}$$

where a, b, and c belong to GF_{p^n} and $(a, b) \neq (0, 0)$. Equation (6.1) is then called the *equation of the line.* If the point (x', y') is contained in the line $ax + by + c = 0$, then

$$ax' + by' + c = 0.$$

If we multiply the equation (6.1) by a nonzero element ρ of the field GF_{p^n}, the equation becomes

$$\rho ax + \rho by + \rho c = 0. \tag{6.2}$$

If the point (x', y') satisfies (6.1) it also satisfies (6.2). Hence, two linear equations that can be obtained from each other on multiplication by a nonzero element of GF_{p^n} represent the same line.

Theorem 6.1. The finite geometry $EG(2, p^n)$ has s^2 points and $s^2 + s$ lines, where $s = p^n$.

Proof. The coordinates of an arbitrary point are (x, y). Since each x and y can assume s values, the number of points is s^2.

The equation of any arbitrary line is $ax + by + c = 0$, where a, b, and c belong to GF_{p^n} and $(a, b) \neq (0, 0)$. We shall consider two cases separately.

Case 1. $b \neq 0$. Dividing by b and transferring the constant term and the term containing x to the other side we can write the equation of the line as

$$y = -\frac{a}{b}x - \frac{c}{d}.$$

Putting $-a/b = m$ and $-c/d = \beta$, we can write the equation of the line as

$$y = mx + \beta. \tag{6.3}$$

Since each m and β can assume $s = p^n$ values, the number of lines belonging to this case is s^2.

Case 2. $b = 0$. Then $a \neq 0$. Dividing by a and transferring the constant term to the other side we can write the equation of the line as

$$x = -\frac{c}{a}.$$

Putting $-c/a = \gamma$, this becomes

$$x = \gamma. \tag{6.4}$$

Since γ can assume s different values the number of lines belonging to this case is s.

Taking both cases together, the required number of lines in $EG(2, p^n)$ is $s^2 + s$.

We cannot define distances and angles in finite geometry.

Example 1. Consider the geometry EG(2, 2) based on the field of GF_2. The field contains only two elements, namely, the classes 0 and 1. Here $s = 2$. There are four points,

$$(0,0),\ (1,0),\ (0,1),\ (1,1).$$

The number of lines is $s^2 + s = 6$. They are shown in Table 6.1, giving the equation of the line and the points contained in it.

Table 6.1

Equation of the Line	Points Contained in the Line
$y = 0$	(0, 0), (1, 0)
$y = 1$	(0, 1), (1, 1)
$y = x$	(0, 0), (1, 1)
$y = x + 1$	(1, 0), (0, 1)
$x = 0$	(0, 0), (0, 1)
$x = 1$	(1, 0), (1, 1)

There is no completely satisfactory way of illustrating a finite plane. We show two different ways of representing EG(2, 2) in Figure 6.1.

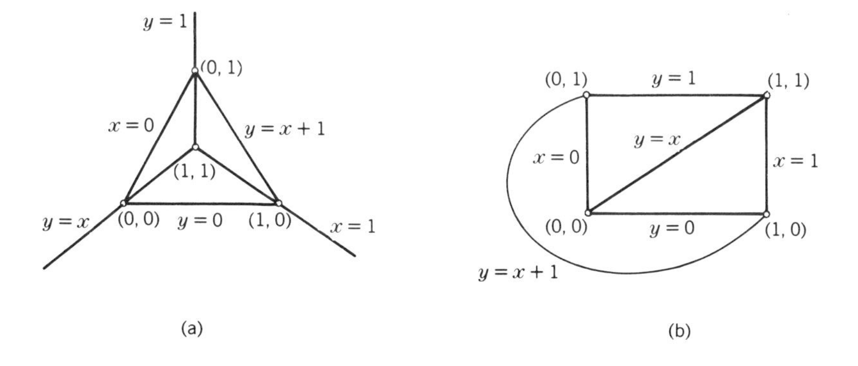

Figure 6.1

Example 2. Consider the geometry EG(2, 3) based on the field GF_3. Now $p^n = 3 = s$. There are $s^2 = 9$ points and $s^2 + s = 12$ lines. The nine points are

$$(0, 0), (0, 1), (0, 2), (1, 0), (1, 1), (1, 2), (2, 0), (2, 1), (2, 2).$$

The twelve lines together with the points on them are shown in Table 6.2.

Table 6.2

Equation of the Line	Points Contained in the Line		
$y = 0$	(0, 0),	(1, 0),	(2, 0)
$y = 1$	(0, 1),	(1, 1),	(2, 1)
$y = 2$	(0, 2),	(1, 2),	(2, 2)
$y = x$	(0, 0),	(1, 1),	(2, 2)
$y = x + 1$	(0, 1),	(1, 2),	(2, 0)
$y = x + 2$	(0, 2),	(1, 0),	(2, 1)
$y = 2x$	(0, 0),	(1, 2),	(2, 1)
$y = 2x + 1$	(0, 1),	(1, 0),	(2, 2)
$y = 2x + 2$	(0, 2),	(1, 1),	(2, 0)
$x = 0$	(0, 0,),	(0, 1),	(0, 2)
$x = 1$	(1, 0),	(1, 1),	(1, 2)
$x = 2$	(2, 0),	(2, 1),	(2, 2)

One representation of the plane EG(2, 3) is given in Figure 6.2a. Another representation is given in Figure 6.2b. Here lines look straight but some of the points appear in two places. Nevertheless, wherever the same pair (x', y') occurs it is to be regarded as the same point.

EXERCISES

1. Consider the geometry EG(2, 2^2).
(a) How many points and lines does this geometry have?
(b) Write equations for all lines in the geometry, and find the points contained in each line.

2. Find an equation for the line containing the points (2, 3) and (1, 2) in EG(2, 5).

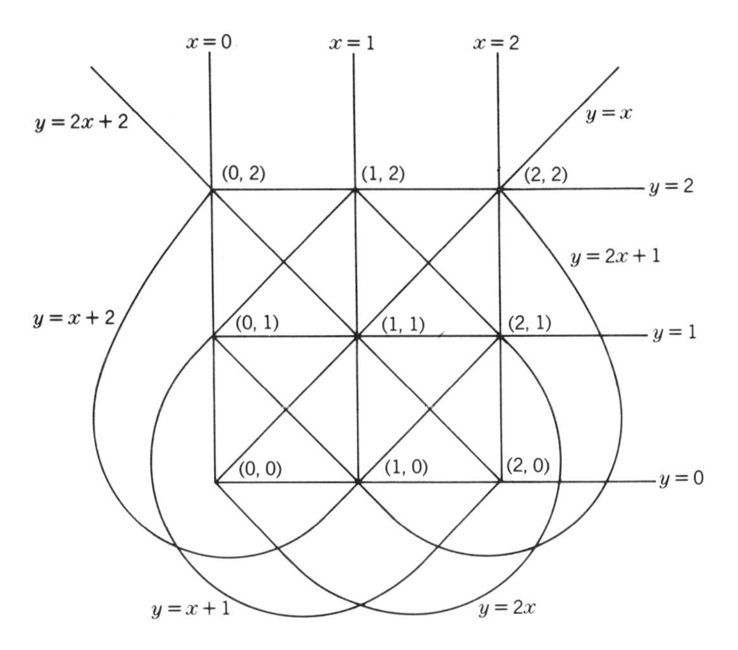

Figure 6.2a

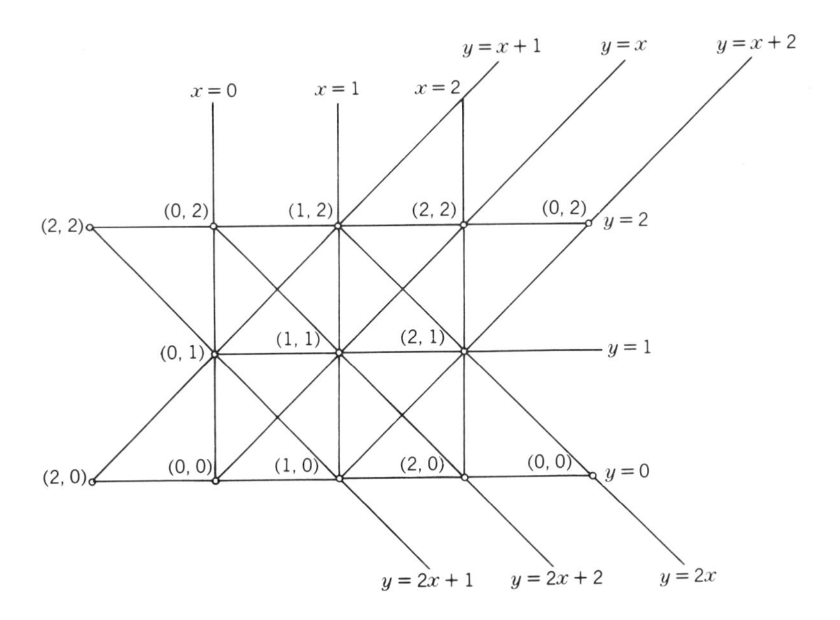

Figure 6.2b

3. Find the intersection of the lines $5x + 2y = 1$ and $3x + 9y = 9$ in EG(2, 11).

4. Write equations for all of the lines in EG(2, 5) that do not have any points in common with the line $2x + 3y = 4$. What would you call these lines?

5. Find the points contained in the following lines of the geometry EG(2, 19):
(a) $13x + 4y = 7$.
(b) $x = 3$.

6.2. SOME PROPERTIES OF THE GEOMETRY EG(2, p^n)

Theorem 6.2. Any two points of EG(2, p^n) are joined by exactly one line.

Proof. Let (x_1, y_1), (x_2, y_2) be two distinct points of EG(2, p^n). The condition that both be on the line

$$ax + by + c = 0$$

is

$$ax_1 + by_1 + c = 0,$$
$$ax_2 + by_2 + c = 0.$$

Eliminating a, b, and c we have

$$\begin{vmatrix} x & y & 1 \\ x_1 & y_1 & 1 \\ x_2 & y_2 & 1 \end{vmatrix} = 0$$

or

$$x(y_1 - y_2) - y(x_1 - x_2) + (x_1 y_2 - y_1 x_2) = 0. \qquad (6.5)$$

This is the equation of the unique line joining the two points (x_1, y_1) and (x_2, y_2).

Lemma 6.1. If $(a_1, b_1) \neq (0, 0)$, $(a_2, b_2) \neq (0, 0)$, and $a_1 b_2 - a_2 b_1 = 0$, where

a_1, a_2, b_1, and b_2 belong to GF_{p^n}, then there exists a nonzero element ρ of GF_{p^n} such that

$$a_2 = \rho a_1, \quad b_2 = \rho b_1.$$

Proof. Since $(a_1, b_1) \neq (0, 0)$ suppose $b_1 \neq 0$. Then from

$$a_1 b_2 - a_2 b_1 = 0$$

we have

$$a_2 = \frac{b_2}{b_1} a_1. \tag{6.6}$$

Then $b_2 \neq 0$, because if $b_2 = 0$ then from (6.6) $a_2 = 0$, which contradicts $(a_2, b_2) \neq (0, 0)$. Hence, taking $\rho = b_2/b_1 \neq 0$ we have

$$a_2 = \rho a_1, \quad b_2 = \rho b_1.$$

Similarly, we can treat the case when $a_1 \neq 0$.

Two distinct lines of EG(2, p^n) are called *parallel* if they have no point in common. A line can be regarded as parallel to itself.

Theorem 6.3. Two distinct lines of EG(2, p^n) are either parallel or intersect in exactly one point.

Proof. Let l_1 and l_2 be two distinct lines of EG(2, p^n) with equations

$$a_1 x + b_1 y + c_1 = 0, \qquad (a_1, b_1) \neq (0, 0),$$
$$a_2 x + b_2 y + c_2 = 0, \qquad (a_2, b_2) \neq (0, 0).$$

The condition for the point (x', y') to be common to the two lines is

$$a_1 x' + b_1 y' + c_1 = 0.$$
$$a_2 x' + b_2 y' + c_2 = 0.$$

Multiplying the first equation by b_2 and the second equation by b_1 and subtracting we have

$$(a_1 b_2 - a_2 b_1) x' + (c_1 b_2 - c_2 b_1) = 0. \tag{6.7}$$

Similarly, multiplying the first equation by a_2 and the second equation by a_1 and subtracting, we have

$$(a_2b_1 - a_1b_2)y' + (a_2c_1 - a_1c_2) = 0. \tag{6.8}$$

Case 1. Suppose $a_1b_2 - a_2b_1 \neq 0$. Then from (6.7) and (6.8) we have

$$x' = \frac{b_1c_2 - b_2c_1}{a_1b_2 - a_2b_1}. \tag{6.9}$$

$$y' = \frac{c_1a_2 - c_2a_1}{a_1b_2 - a_2b_1}. \tag{6.10}$$

Hence, in this case the two lines intersect in exactly one point (x', y'), where x' and y' are given by (6.9) and (6.10).

Case 2. Suppose $a_1b_2 - a_2b_1 = 0$. Then from Lemma 6.1 there exists a nonzero element ρ of GF_{p^n} such that $a_2 = \rho a_1$, $b_2 = \rho b_1$. Hence, from (6.7) and (6.8) we have

$$b_1(\rho c_1 - c_2) = 0,$$
$$a_1(\rho c_1 - c_2) = 0.$$

Now $c_2 \neq \rho c_1$, otherwise the two lines coincide. Hence, $a_1 = 0$, $b_1 = 0$, which is a contradiction. Thus in this case there cannot exist a common point (x', y'), that is, the two lines are parallel.

Corollary 6.1. The equation of any line parallel to the line l with equation $ax + by + c = 0$ may be taken as $ax + by + c' = 0$.

Proof. Let l' be any line parallel to l. If its equation is

$$a'x + b'y + c_0 = 0, \tag{6.11}$$

then $a'b - ab' = 0$. Hence, from Lemma 6.1 $a' = \rho a$, $b' = \rho b$, $\rho \neq 0$. Substituting in (6.11) and dividing by ρ the equation of l' becomes

$$ax + by + \frac{c_0}{\rho} = 0.$$

Putting $c_0/\rho = c'$, the equation reduces to the required form.

Theorem 6.4. There are exactly s points on each line of EG(2, p^n), where $s = p^n$.

Proof. We have seen that the equation of a line can be written either as

$$\text{(i)} \quad y = mx + \beta$$

or as

$$\text{(ii)} \quad x = \gamma.$$

In case (i), corresponding to each value of x, there is exactly one value of y. Since x can takc s valucs, wc have exactly s points on the line.

In case (ii), x has a fixed value γ but y can be chosen in s different ways. Hence there are s points on the line.

Theorem 6.5. From any point P there passes exactly one line parallel to a given line l.

Proof. Let P be the point (x', y') and l the line $ax + by + c = 0$. Any line m parallel to l has an equation of the form $ax + by + c' = 0$, from Corollary 6.1. If this line passes through P then

$$ax' + by' + c' = 0.$$

Hence $c' = -(ax' + by')$. Substituting for c' the equation of m becomes

$$ax + by - (ax' + by') = 0. \tag{6.12}$$

Thus, there is a unique line through P parallel to l and its equation is (6.12).

Theorem 6.6. If two lines m_1 and m_2 are each parallel to the line l, then they are parallel to one another.

Proof. Let the equation of l be $ax + by + c = 0$. Then the equations of m_1 and m_2 must be of the form $ax + by + c_1 = 0$ and $ax + by + c_2 = 0$. They are clearly parallel to one another.

Theorem 6.7. The $s^2 + s$ lines of EG(2, p^n) can be divided into $s + 1$ sets, each consisting of s lines, such that any two lines of the same set are parallel to one another. Each set is called a *parallel pencil.* Through any point there passes exactly one line of a given parallel pencil.

Proof. We have seen that the equation of a line can be taken in one of two forms:

$$\text{(i)}\quad y = mx + \beta,$$
$$\text{(ii)}\quad x = \gamma.$$

For a fixed m there are s lines of the form (i) obtained by allowing β to assume all possible values in GF_{p^n}. These s lines are mutually parallel to each other. Through any point (x', y') there passes exactly one of these lines, namely the line $y = mx + (y' - mx')$. They thus form a parallel pencil. The slope of the parallel pencil so defined is called m. To each value of m there corresponds in this way a parallel pencil. We thus have s parallel pencils of type (i).

Again there are s lines of the form (ii) obtained by allowing γ to assume all possible values in GF_{p^n}. These lines are mutually parallel to one another. Through any point (x', y') there passes exactly one of these lines, namely the line $x = x'$. They thus form a parallel pencil. The slope of the parallel pencil so defined is said to be ∞. Thus there is a single parallel pencil of type (ii).

Altogether, we get $s + 1$ parallel pencils, s belonging to type (i), and one belonging to type (ii). Clearly each line must belong to exactly one parallel pencil.

Corollary 6.2 Through any point P of EG(2, p^n) there pass exactly $s + 1$ lines, one belonging to each parallel pencil.

Since the symbol x is used to denote the x-coordinate of a point, in order to avoid confusion we shall take the polynomials denoting elements of GF_{p^n} to be polynomials in variable t instead of x. Thus, in the following examples the elements of the fields GF_{3^2}, GF_{2^2}, and GF_{2^3} will be taken exactly as in Section 5.6, except that instead of taking powers of x and polynomials in x we shall take powers of t and polynomials in t.

Example 3. Consider the geometry EG(2, 2^2). The elements of GF_{2^2} may be taken in the multiplicative form 0, 1, t, t^2. There are 16 points and 20 lines.

(*i*) Find the equation of the line joining the points (1, t) and (t, t^2). Using the formula (6.5) the required equation is

$$(t - t^2)x - (1 - t)y + (t^2 - t^2) = 0$$

or

$$x + t^2y = 0.$$

(*ii*) Determine whether the following lines are parallel. If not, find their point of intersection

$$x + ty + t^2 = 0, \quad t^2x + y + 1 = 0.$$

Do the same for the pair of lines

$$x + ty + t = 0, \quad tx + ty + 1 = 0.$$

For the first pair, $ab' - a'b = 1 - t^3 = 1 + 1 = 0$. Hence, they are parallel.

For the second pair, $ab' - a'b = t - t^2 = t + t^2 = 1$. Hence, they are not parallel. If their intersection is the point (x', y') then

$$x' + ty' + t = 0,$$
$$tx' + ty' + 1 = 0.$$

Solving these equations we get

$$(x', y') = (1, t).$$

Example 4. Consider the geometry EG(2, 3^2). The elements of GF_{3^2} can be taken in the multiplicative form 0, 1, t, t^2, t^3, t^4, t^5, t^6, and t^7

Suppose we wish to find the equation of the line parallel to $x + t^3y + t^5 = 0$ and passing through the point (t^6, t^2).

Using formula (6.12) the required equation is

$$x + t^3y - (t^6 + t^5) = 0.$$

But

$$-(t^6 + t^5) = -(t + 2 + 2t) = -2 = 1.$$

Hence, the required equation is

$$x + t^3y + 1 = 0.$$

EXERCISES

1. For the geometry EG(2, 2^3):
(a) Find all points on the line $x + t^3y = t^5$.

(b) Find the equation of the line passing through the point $(1, t^6)$ parallel to the line $y = t^5x + t^3$.

(c) Find the point of intersection of the lines

$$t^5x + t^3y + 1 = 0,$$
$$t^2x + t^4y + t^7 = 0.$$

(d) Find the equation of the line joining $(1, t^5)$ and $(t^6, 0)$.
[See Section 5.6 for a description of the field $GF(2^3)$. Here we are using $1, t, t^2$, and so on, as the multiplicative representation of the elements of GF_{2^3}.]

2. Find all lines through the point (t^2, t^3) in the geometry $EG(2, 3^2)$.

3. In the geometry $EG(2, 7)$ based on the field GF_7 answer the following questions:

(a) Find the equation of the line joining the points (3, 4) and (2, 6).

(b) Find the point of intersection of the lines

$$3x + 4y = 2 \quad \text{and} \quad 2x + 3y = 6.$$

(c) Find the equation of the line parallel to $3x + 4y = 5$ and passing through the point (2, 3).

6.3. THE PROJECTIVE GEOMETRY PG(2, p^n)

In the Euclidean finite geometry of two dimensions $EG(2, p^n)$ the relation between points and lines is not symmetrical, for two points are always joined by a line, but two lines may or may not intersect in a point. We shall now discuss a more symmetrical geometry in which not only two distinct points are joined by a unique line, but two distinct lines intersect in a unique point. This geometry is called the *projective geometry* of two dimensions based on the field GF_{p^n} and is denoted by $PG(2, p^n)$.

This geometry is obtained by extending the Euclidean geometry in the following manner: Since we want any two lines to intersect, we create one new point corresponding to each parallel pencil and adjoin it to each line of the pencil. The new points are called *points at infinity*. The already existing points of $EG(2, p^n)$ are called *finite points* to distinguish them from the points at infinity. We also create a new line called the *line at infinity*, which contains all the points at infinity but contains no finite points. The already existing lines of $EG(2, p^n)$ are called *finite lines*. The extended geometry so obtained is the geometry $PG(2, p^n)$.

Theorem 6.8. In the projective geometry PG(2, p^n) there are $s^2 + s + 1$ points and $s^2 + s + 1$ lines. Each line contains $s + 1$ points, and each point is contained in $s + 1$ lines. Any two distinct points are joined by a unique line and any two distinct lines intersect in a unique point.

Proof. There are s^2 finite points and $s + 1$ points at infinity (one corresponding to each parallel pencil). Hence there are $s^2 + s + 1$ points in the extended geometry.

There are $s^2 + s$ finite lines and one line at infinity. Hence there are $s^2 + s + 1$ lines.

For any fixed $m \in GF_{p^n}$ we have a parallel pencil [of type (i)] with slope m. Any line of this pencil has an equation $y = mx + \beta$, where m is the same for each line of the pencil but β differs for different lines of the pencil. The point at infinity corresponding to this pencil may be coordinatized by (m).

Again there is a single parallel pencil of type (ii) with slope ∞. The corresponding point may be coordinatized by (∞).

The line at infinity is denoted by l_∞. Now l_∞ contains the $s + 1$ points at infinity and no finite points. Any finite line contains s finite points and the point at infinity corresponding to the pencil to which it belongs. Thus the line $y = mx + \beta$ contains the point (m) at infinity and the line $x = \gamma$ contains the point (∞). Thus each line of the extended geometry contains $s + 1$ points.

Again each finite point is contained in $s + 1$ finite lines, one belonging to each parallel pencil. A point at infinity is contained in each of the s lines of the parallel pencil to which it corresponds and is also contained in the line at infinity. Thus each point is contained in exactly $s + 1$ lines.

Two finite points are joined by a unique (finite) line. The points at infinity lie on l_∞. Next consider a finite point P and a point at infinity Q_∞. P lies on exactly one line belonging to the parallel pencil to which Q_∞ corresponds. This is the unique line joining P and Q_∞.

Two nonparallel finite lines intersect in a unique finite point. Two parallel finite lines intersect in the point at infinity corresponding to the parallel pencil to which both belong. The line at infinity and a finite line intersect in a point at infinity that corresponds to the parallel pencil to which the finite line belongs. Thus, two distinct lines always intersect in a unique point.

Example 5. Consider the geometry PG(2, 2). This is obtained by extending EG(2, 2).

In Table 6.1 the horizontal lines divide the six lines of EG(2, 2) into three parallel pencils. To each line of a given parallel pencil we have to adjoin a new point at infinity and then take a new line containing all the points at infinity. Thus the geometry PG(2, 2) contains seven points and seven lines, shown in Table 6.3.

Table 6.3

Equation of the Line	Points Contained in the Line
$y = 0$	(0,0), (1,0), (0)
$y = 1$	(0,1), (1,1), (0)
$y = x$	(0,0), (1,1), (1)
$y = x + 1$	(1,0), (0,1), (1)
$x = 0$	(0,0), (0,1), (∞)
$x = 1$	(1,0), (1,1), (∞)
l_∞	(0), (1), (∞)

We may draw a diagram of the plane PG(2, 2) by extending the lines in Figure 6.1(a) and obtain a representation of PG(2, 2) given in Figure 6.3. The plane PG(2, 2) is known as the Fano plane.

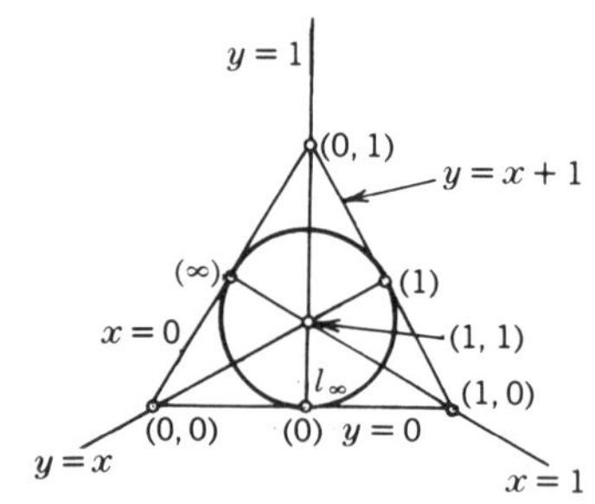

Figure 6.3

Example 6. Consider the geometry PG(2, 3). This is obtained by extending EG(2, 3).

In Table 6.2 the horizontal lines divide the twelve lines of EG(2, 3) into four parallel pencils. The points to be adjoined to the lines of various parallel pencils are shown below.

Parallel Pencil	Point at Infinity To Be Adjoined
$y = \beta, \beta = 0,1,2$	(0)
$y = x + \beta, \beta = 0,1,2$	(1)
$y = 2x + \beta, \beta = 0,1,2$	(2)
$x = \gamma, \gamma = 0,1,2$	(∞).

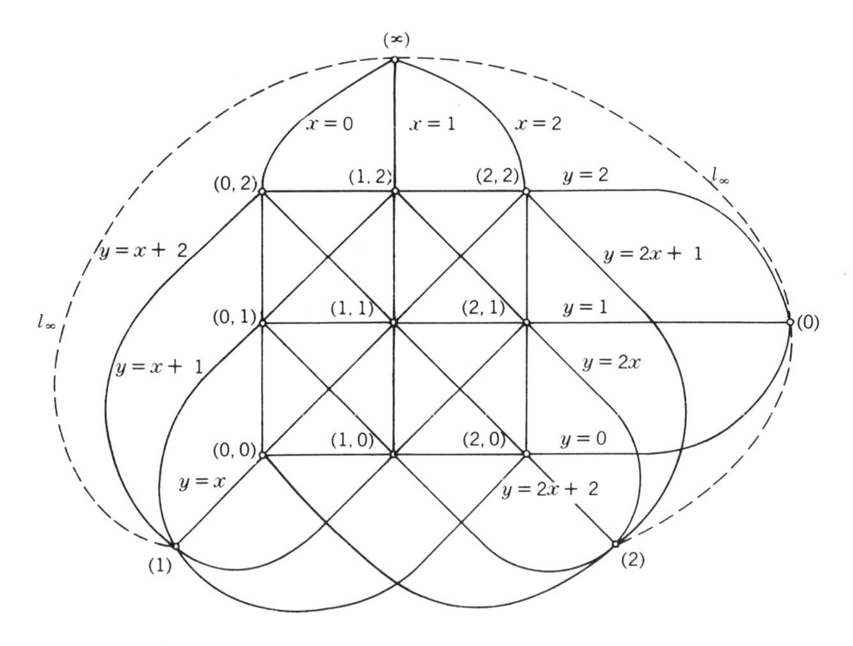

Figure 6.4

We also take a new line containing the four points at infinity. Thus PG(2, 3) contains thirteen points and thirteen lines. A representation obtained by extending the lines in Figure 6.2a is shown in Figure 6.4.

EXERCISES

1. List the lines and points of the geometry PG(2, 2^2).
2. Find the equation of the line through the points $(t, 1)$ and $(t + 1)$ in PG(2, 2^2).
3. Find all points on the line $y = t^3x + t^2$ in the geometry PG(2, 2^3). [Here we use the notation from Section 5.6 for GF_{2^3}, with t^3 for x^3, etc.]
4. Find the point of intersection of the lines $t^3y + t^2x + t = 0$ and $y = t^3x + t^2$ in PG(2, 3^2). [GF_{3^2} as in Section 5.6.]
5. The geometry EG(2, 2^3) is extended to PG(2, 2^3) by the addition of the line at infinity. Answer the following questions with respect to PG(2, 2^3).
   (a) Find the equation of the line joining the points (t^2, t^5) and (t^3).

(b) Find the equation of the line joining the points (t^2, t^5) and (∞).

(c) Find the point of intersection of the lines $t^5x + t^3y + 1 = 0$ and $t^6x + t^4y + t^2 = 0$.

(d) Find the point of intersection of the lines $t^5x + t^3y + 1 = 0$ and $t^6x + t^3y + t^2 = 0$.

6.4. REMARKS

We have seen that a theory very like ordinary analytic geometry can be developed using finite fields instead of the real numbers. Kirkman pioneered the study of finite structures in 1847 when he proved the existence of certain triple systems (equivalent to projective planes), which we will define in Chapter 8. Finite geometries were first considered as such by Fano in 1892. His geometry was 3-dimensional, containing 15 points, 35 lines, and 15 planes, but each plane was our familiar Fano plane. The theory was developed in general in the early part of this century by Veblen and Bussey. In particular, they showed how to construct finite projective planes of order p^n for every prime p and positive integer n.

Everything that you ever might want to know about finite geometries is contained in the classic book by Dembowski [1], although it is sometimes hard to read. A more readable treatment, which ties our Chapter 6, 7, and 8 together in an interesting way, is by Cameron and Van Lint [2].

[1] P. Dembowski, *Finite Geometries*, Springer-Verlag, New York, 1968.

[2] P. J. Cameron and J. H. Van Lint, *Graph Theory, Coding Theory and Block Designs*, Cambridge University Press, Cambridge, England, 1975.

CHAPTER 7

Orthogonal Latin Squares and Error Correcting Codes

7.1. LATIN SQUARES

A *Latin square* of order s is identified as an $s \times s$ square, the s^2 cells of which are occupied by s distinct symbols (which may be Latin or Greek letters or Arabic numerals) such that each symbol occurs once in each row and once in each column. For example, the square exhibited in Figure 7.1 is a Latin square of order 6 whose cells are occupied by the symbols A, B, C, D, E, F.

B D A E C F
D A C B F E
E F D A B C
A C E F D B
C B F D E A
F E B C A D

Figure 7.1

Two Latin squares of the same order are said to be *orthogonal* if, on superposition, each symbol of the first square occurs exactly once with each symbol of the second square. Figure 7.2 shows two orthogonal Latin squares of order 5, the symbols for the squares being 0, 1, 2, 3, 4. The term Latin square derives from the fact that Euler used Latin letters as the symbols for the first square in any pair of orthogonal Latin squares.

2 4 1 0 3	4 0 2 1 3
4 1 0 3 2	2 1 3 4 0
1 0 3 2 4	3 4 0 2 1
0 3 2 4 1	0 2 1 3 4
3 2 4 1 0	1 3 4 0 2

Figure 7.2

A set of Latin squares (all of the same order), any two of which are orthogonal, is said to be *a set of mutually orthogonal Latin squares.*

A Latin square is said to be in the *standard form* if the symbols in the initial row are in the natural order. Thus when the symbols are the integers $0, 1, 2, \ldots, s - 1$ the square is in the standard form if the initial row is $0, 1, 2, \ldots, s - 1$.

A Latin square can always be brought to the standard form by renaming the symbols. If two Latin squares are orthogonal, the renaming can be done independently for each square without destroying orthogonality. For example, to standardize the first square in Figure 7.2 we make the transformation

$$2 \to 0, \quad 4 \to 1, \quad 1 \to 2, \quad 0 \to 3, \quad 3 \to 4.$$

Similarly, to standardize the second square, we make the transformation

$$4 \to 0, \quad 0 \to 1, \quad 2 \to 2, \quad 1 \to 3, \quad 3 \to 4.$$

The transformed squares are shown in Figure 7.3 and are orthogonal to one another.

```
0 1 2 3 4      0 1 2 3 4
1 2 3 4 0      2 3 4 0 1
2 3 4 0 1      4 0 1 2 3
3 4 0 1 2      1 2 3 4 0
4 0 1 2 3      3 4 0 1 2
```

Figure 7.3

Given any Latin square of order s, we shall number the successive rows (columns) of the square as $0, 1, 2, \ldots, s - 1$. Thus the initial row is the row 0, the next row is the row 1, and so on; the same holds for columns. The cell in the ith row and jth column will be called the cell (i, j).

There are exactly two Latin squares of order 2. Once the cell $(0, 0)$ is filled the symbols in the other cells are uniquely determined. Thus if the symbols are 0 and 1, then the two possible Latin squares of order 2 are

```
0 1    1 0
1 0    0 1
```

Next consider Latin squares of order 3. We shall show that once the initial row and column are filled, there is one way to complete the Latin square. Suppose we take the initial row and column in the natural order and start with

0 1 2
1 . .
2 . .

We cannot put 1 in the cell (1, 1), for this would mean two 1's in the row number 1. If we put 0 in the cell (1, 1) then the symbol in the cell (1, 2) must be 2 in order to have all three symbols in the row number 1. But now we have two 2's in the column number 2. Hence the cell (1, 1) must have the symbol 2. Now there is only one way to complete the square. Thus we get

0 1 2
1 2 0
2 0 1

We can permute the columns in six different ways. Corresponding to each of these permutations there are two ways to permute the rows other than the initial row. Hence there are twelve distinct Latin squares of order 3. They are shown in Figure 7.4.

0 1 2   0 2 1   1 2 0   1 0 2   2 0 1   2 1 0
1 2 0   1 0 2   2 0 1   2 1 0   0 1 2   0 2 1
2 0 1   2 1 0   0 1 2   0 2 1   1 2 0   1 0 2

0 1 2   0 2 1   1 2 0   1 0 2   2 0 1   2 1 0
2 0 1   2 1 0   0 1 2   0 2 1   1 2 0   1 0 2
1 2 0   1 0 2   2 0 1   2 1 0   0 1 2   0 2 1

Figure 7.4

There are exactly four Latin squares of order 4 in which the initial row and the initial column are in the natural order 0, 1, 2, 3, because once these are filled the cell (1, 1) must be filled with 2, 3, or 0. There is now exactly one way to complete the square if the cell (1, 1) contains either the symbol 2 or the symbol 3, but there are two ways to complete the square if the cell (1, 1) contains the symbol 0. These four Latin squares are shown in Figure 7.5.

0 1 2 3   0 1 2 3
1 2 3 0   1 3 0 2
2 3 0 1   2 0 3 1
3 0 1 2   3 2 1 0

0 1 2 3   0 1 2 3
1 0 3 2   1 0 3 2
2 3 0 1   2 3 1 0
3 2 1 0   3 2 0 1

Figure 7.5

From any square in Figure 7.5 we can obtain 144 squares by first permuting the four columns in all 24 possible ways and then permuting the three rows other than the initial row in all 6 possible ways. Thus there are altogether $4 \cdot 144 = 576$ distinct Latin squares of order 4.

There are 56 different Latin squares of order 5 in which the initial row and column are in the natural order, each giving rise to $5! \cdot 4! = 2{,}880$ different squares by permuting the columns and then the rows other than the initial row. Thus there are 161,280 distinct Latin squares of order 5.

Theorem 7.1. There cannot exist a set of more than $s - 1$ mutually orthogonal Latin squares of order s.

Proof. Suppose there exists a set of m mutually orthogonal Latin squares of order s. By renaming the symbols, we can transform each square to the standard form in which the initial row is occupied by the symbols $0, 1, \ldots, s - 1$ in order. Thus, in each square, the cell $(0, j)$ contains the symbol j, where $0 \leqslant j \leqslant s - 1$. The standardized squares are mutually orthogonal. Since the cell $(0, 0)$ contains the symbol 0, the symbol in the cell $(1, 0)$ must be different from 0 for each of the m standardized squares. When the uth square is superimposed on the tth square, the symbol j of the uth square occurs together with the symbol j of the tth square in the cell $(0, j)$; $j = 0, 1, \ldots, s - 1$. Hence the symbols in the cell $(1, 0)$ of these two squares must be different. Thus, the cells $(1, 0)$ of the m standardized squares are occupied by different nonzero symbols. Since there are only $s - 1$ nonzero symbols, $m \leqslant s - 1$.

EXERCISES

1. Standardize the Latin square

B D A E C F
D A C B F E
E F D A B C
A C E F D B
C B F D E A
F E B C A D

2. Two of the following squares are orthogonal. Which two?

(a)	(b)
1 0 3 2 4	2 4 1 0 3
4 1 0 3 2	1 0 3 2 4
0 3 2 4 1	4 1 0 3 2
2 4 1 0 3	0 3 2 4 1
3 2 4 1 0	3 2 4 1 0

(c)	(d)
4 0 2 1 3	3 4 0 2 1
2 1 3 4 0	2 1 3 4 0
3 4 0 2 1	0 2 1 3 4
1 3 4 0 2	4 0 2 1 3
0 2 1 3 4	1 3 4 0 2

3. Show that there are at least three distinct ways of completing the following square so that we get a Latin square.

A	B	C	D	E
B	A	D	·	·
C	·	·	·	·
D	·	·	·	·
E	·	·	·	·

4. Show that there exists no square orthogonal to

0 1 2 3
3 0 1 2
2 3 0 1
1 2 3 0

7.2. COMPLETE SETS OF ORTHOGONAL LATIN SQUARES

A set of $s - 1$ mutually orthogonal Latin squares of order s is said to be a *complete set of mutually orthogonal Latin squares.* We shall show that when s is a prime power, we can obtain a complete set of mutually orthogonal Latin squares of order s. There is no known example of a complete set of $s - 1$ mutually orthogonal Latin squares of order s when s is not a prime power, though the existence of such a set has not been disproved except for certain special values of s. The smallest value of s for which the question is open is 10.

Let s be a prime power. Then there exists a finite field GF_{p^n} with $s = p^n$ elements. Let these elements be

$$\alpha_0 = 0, \alpha_1, \ldots, \alpha_{s-1}.$$

Take an $s \times s$ square and in the cell (i, j) of this square put the integer $u = u(i, j)$ given by

$$\alpha_u = \alpha_i \alpha_t + \alpha_j, \tag{7.1}$$

where α_t is a fixed nonzero element of GF_{p^n}. We shall show that we obtain in this way a Latin square $[L_t]$. To prove this we have to show that each row and each column contains the symbols $0, 1, 2, \ldots, s - 1$ exactly once. In the row i the symbol u occurs in the column j given by

$$\alpha_j = \alpha_u - \alpha_i\alpha_t.$$

In the column j the symbol u occurs in the row i given by

$$\alpha_i = \frac{\alpha_u - \alpha_j}{\alpha_t}.$$

This proves our claim. In formula (7.1), α_t can take any one of the values $\alpha_1, \alpha_2, \ldots, \alpha_{s-1}$. Corresponding to each of these values we get a Latin square. We thus get $s - 1$ Latin squares

$$[L_1], [L_2], \ldots, [L_{s-1}].$$

We claim that any two of these Latin squares are mutually orthogonal. Let $[L_t]$ and $[L_t']$, $1 \leqslant t \leqslant s - 1$, $1 \leqslant t' \leqslant s - 1$, $t \neq t'$ be two of these Latin squares. When superposed the symbol u of the first square occurs together with the symbol u' of the second square in the cell (i, j) if and only if

$$\alpha_u = \alpha_i\alpha_t + \alpha_j,$$

$$\alpha_{u'} = \alpha_i\alpha_{t'} + \alpha_j.$$

Solving these two equations for α_i and α_j we obtain

$$\alpha_i = \frac{\alpha_u - \alpha_{u'}}{\alpha_t - \alpha_{t'}}, \qquad \alpha_j = \frac{\alpha_t\alpha_{u'} - \alpha_{t'}\alpha_u}{\alpha_t - \alpha_{t'}}.$$

Hence the cell (i, j) is uniquely determined. This substantiates our claim. We have thus obtained a complete set of mutually orthogonal Latin squares of order s, for any prime power s. Thus we have the following theorem.

Theorem 7.2. For any given prime power s there exists a complete set of mutually orthogonal Latin squares of order s.

Of course, here we have proven the existence of one complete set of order $s = p^n$. In general there are many nonisomorphic sets of a given prime power order. Equation (7.1) may be called the equation of the Latin square $[L_t]$.

Further simplification is introduced by a judicious identification of the nonzero elements of GF_{p^n} with $\alpha_1, \alpha_2, \ldots, \alpha_{s-1}$. In the special case when $n = 1$ (*i.e.*, $s = p$ is a prime number), we can identify α_i with the residue class (i), (mod p). Then the number u to be put in the cell (i, j) of $[L_t]$ is given by

$$u = it + j, \tag{7.2}$$

where i, t, j, and u are standard representatives of residue classes (mod p).

Example 1. Let $s = 5$; then the four mutually orthogonal Latin squares of order 5 obtained by the above method are exhibited in Figure 7.6.

$[L_1]$ $u = i + j$

0	1	2	3	4
1	2	3	4	0
2	3	4	0	1
3	4	0	1	2
4	0	1	2	3

$[L_2]$ $u = 2i + j$

0	1	2	3	4
2	3	4	0	1
4	0	1	2	3
1	2	3	4	0
3	4	0	1	2

$[L_3]$ $u = 3i + j$

0	1	2	3	4
3	4	0	1	2
1	2	3	4	0
4	0	1	2	3
2	3	4	0	1

$[L_4]$ $u = 4i + j$

0	1	2	3	4
4	0	1	2	3
3	4	0	1	2
2	3	4	0	1
1	2	3	4	0

Figure 7.6

In the general case $s = p^n$ where n may be equal to or greater than 1, a useful identification is provided by taking

$$\alpha_1 = 1, \quad \alpha_2 = x, \ldots, \quad \alpha_i = x^{i-1}, \ldots, \quad \alpha_{s-1} = x^{s-2}, \tag{7.3}$$

where x is a primitive element of GF_{p^n} corresponding to a primitive root. When $n = 1$, x is the residue class of the prime p.

Example 2. Let $s = 4$. We then take the elements of GF_{2^2} as

$$\alpha_0 = 0, \quad \alpha_1 = 1, \quad \alpha_2 = x, \quad \alpha_3 = x^2 = 1 + x$$

as in Section 5.6(b). The three mutually orthogonal Latin squares $[L_1]$, $[L_2]$, $[L_3]$ are shown in Figure 7.7, where the equation of each square is given at the top.

$[L_1]$	$[L_2]$	$[L_3]$
$\alpha_u = \alpha_i + \alpha_j$	$\alpha_u = \alpha_i\alpha_2 + \alpha_j$	$\alpha_u = \alpha_i\alpha_3 + \alpha_j$
0 1 2 3	0 1 2 3	0 1 2 3
1 0 3 2	2 3 0 1	3 2 1 0
2 3 0 1	3 2 1 0	1 0 3 2
3 2 1 0	1 0 3 2	2 3 0 1

Figure 7.7

We note that the rows of $[L_2]$ (other than the row 0) are obtainable from the rows of $[L_1]$ by a cyclic permutation, the rows 1, 2, 3 of $[L_2]$ being the same as the rows 2, 3, 1 of $[L_1]$. Again, the rows of $[L_3]$ can be obtained from the rows of $[L_2]$ by the same cyclic permutation. We shall show that this is true in general for sets of orthogonal Latin squares obtained by using the formula (7.1) when the elements $\alpha_0, \alpha_1, \ldots, \alpha_{s-1}$ of GF_{p^n} are chosen as in (7.3).

The element in the cell (i, j) of $[L_t]$ is the number u given by

$$\alpha_u = \alpha_i\alpha_t + \alpha_j, \qquad 0 \leqslant u \leqslant s - 1.$$

The element in the cell $(i - 1, j)$ of $[L_{t+1}]$ is the number u' given by

$$\alpha_{u'} = \alpha_{i-1}\alpha_{t+1} + \alpha_j.$$

Let $1 \leqslant t \leqslant s - 2$, $2 \leqslant i \leqslant s - 1$. Then

$$\alpha_u = x^{i-1}x^{t-1} + \alpha_j = x^{i-2}x^t + \alpha_j = \alpha_{u'},$$

or $u = u'$. Hence the cell $(i - 1, j)$ of $[L_{t+1}]$ contains the same number as the cell (i, j) of $[L_t]$.

Again, the element in the cell $(1, j)$ of $[L_t]$ is the number v given by

$$\alpha_v = \alpha_1\alpha_t + \alpha_j = x^{t-1} + \alpha_j,$$

and the number in the cell $(s - 1, j)$ of $[L_{t+1}]$, $1 \leqslant t \leqslant s - 2$, is the number v' given by

$$\alpha_{v'} = \alpha_{s-1}\alpha_{t+1} + \alpha_j = x^{s+t-2} + \alpha_j = x^{t-1} + \alpha_j,$$

since $x^{s-1} = 1$. Hence the cell $(s - 1, j)$ of $[L_{t+1}]$ contains the same number as the cell $(1, j)$ of $[L_t]$.

We have thus shown that for $1 \leqslant t \leqslant s - 2$, the rows $1, 2, \ldots, s - 1$ of $[L_t]$ are identical with the rows $s - 1, 1, \ldots, s - 2$ of $[L_{t+1}]$.

Example 3. Let $s = 9$. The elements of GF_{3^2} have been obtained in Section 5.6(a). The first Latin square $[L_1]$ together with its equation is shown in Figure 7.8.

$[L_1]$

$\alpha_u = \alpha_i + \alpha_j$

0	1	2	3	4	5	6	7	8
1	5	8	4	6	0	3	2	7
2	8	6	1	5	7	0	4	3
3	4	1	7	2	6	8	0	5
4	6	5	2	8	3	7	1	0
5	0	7	6	3	1	4	8	2
6	3	0	8	7	4	2	5	1
7	2	4	0	1	8	5	3	6
8	7	3	5	0	2	1	6	4

Figure 7.8

The rows of $[L_2]$ can now be obtained by keeping the row 0 unchanged and applying a cyclic permutation to the rows 1, 2, ..., $s - 1$ of the previous square $[L_1]$. In the same way, we can write down the successive squares $[L_3]$, $[L_4]$, ..., $[L_8]$.

Example 4. Let $s = 8$. The elements of GF_{2^3} have been obtained in Section 5.6(c). The first Latin square $[L_1]$ together with its equation is shown in Figure 7.9. The other squares can be obtained from $[L_1]$ in the manner explained before.

$[L_1]$

$\alpha_u = \alpha_i + \alpha_j$

0	1	2	3	4	5	6	7
1	0	6	4	3	7	2	5
2	6	0	7	5	4	1	3
3	4	7	0	1	6	5	2
4	3	5	1	0	2	7	6
5	7	4	6	2	0	3	1
6	2	1	5	7	3	0	4
7	5	3	2	6	1	4	0

Figure 7.9

EXERCISES

1. A complete set of six orthogonal Latin squares of order 7 is obtained by using $u = it + j$ as the equation of the square $[L_t]$, where u, i, t are residue classes (mod 7). The rows and columns are numbered 0, 1, 2, 3, 4, 5, 6.

(a) Write down the row 3 of $[L_5]$.

(b) In which cell does the symbol 4 of $[L_3]$ come together with the symbol 6 of $[L_4]$, when $[L_3]$ and $[L_4]$ are superimposed?

2. A complete set of orthogonal Latin squares of order 9 is constructed as in Example 3.

(a) Find the element in cell (3, 4) of $[L_5]$.

(b) In which square does the symbol 5 occur in row 2 and column 4?

3. Find the 5×5 Latin squares L_1, L_2, L_3, and L_4 of Example 1, construct the geometry EG(2, 5).

4. If we construct eight squares L^k from $L^k = [a_{ij}^k]$, where $a_{ij}^k = i + jk$ (mod 9), which of the squares L^1 to L^8 are Latin? Note this is a case where equation (7.2) is being used, but (7.1) really applies.

7.3. ERROR-CORRECTING CODES

For most people, code and cypher are words that evoke general thoughts of hidden messages and spies. The two words, in fact, have rather different meanings. A cypher is a rule for changing an alphabet into a different form (a rearranged alphabet, numbers, or other symbols), so that a message can be hidden from a casual reader. A code is a list of words or symbols that replace specific other words. Thus, a Russian word written in our alphabet is encyphered, but if it is translated into English, it is encoded. A Russian–English dictionary can be considered a code book—each Russian word has an (English) equivalent in the code specified by the dictionary.

In this section we will be concerned with codes that are used for ensuring clarity of a message, rather than hiding it from view. When words are transmitted bit by bit, as in radio messages through space, some bits may be lost. When the word SHIP is transmitted, interference on one letter may result in receipt of SHLP. The person receiving the message will then have to decide whether the word sent was SHIP or SHOP (usually this will be clear from context). The richness of English makes it very hard to avoid words that are "near" each other in this sense. Thus, it is desirable to encode the message, making each word equivalent to a "word" in a new language, in which no two words are "close." The construction of such languages is the subject of this section.

We use as our alphabet the q symbols 0, 1, 2, ..., $q - 1$. An ordered n-tuple of symbols is an *n-word*. The *distance d* between two words x and y is the number of spots at which the two words differ. Thus (0, 1, 0, 2) and (0, 1, 2, 0) are at distance 2, while (0, 1, 0, 2) and (0, 2, 0, 2) are at distance 1. We claim that the distance so defined is a metric, that is,

(i) $d(x, x) = 0$,
(ii) $d(x, y) = d(y, x)$,
(iii) $d(x, y) + d(y, z) \geqslant d(x, z)$.

The truth of **(i)** and **(ii)** is evident. We shall prove **(iii)**.

Let the ith coordinates of x, y, z be x_i, y_i, z_i, respectively.

If $x_i = z_i$, then either $x_i = y_i = z_i$ or $x_i \neq y_i \neq z_i$. The contribution of the ith coordinate to $d(x, z)$ is 0, but the contribution of the ith coordinates to $d(x, y) + d(y, z)$ is 0 or 2.

If $x_i \neq z_i$, then either $x_i = y_i \neq z_i$ or $x_i \neq y_i = z_i$ or $x_i \neq y_i \neq z_i$. The contribution of the ith coordinate to $d(x, z)$ is 1, but the contribution of ith coordinate to $d(x, y) + d(y, z)$ is 1 or 2.

Thus in every case the contribution of the ith coordinate to $d(x, y) + d(y, z)$ is greater than or equal to the contribution to $d(x, z)$. Hence

$$d(x, y) + d(y, z) \geqslant d(x, z). \tag{7.4}$$

A *code* is a collection of words of length n over the given alphabet of q symbols. The *distance* of the code is the minimum distance between all pairs of distinct code words. Clearly at the same time q should be kept small, to make transmission simple, and for a given n the number of different code words should be as large as possible, to allow a variety of messages. The careful construction of "optimal" codes is an important topic of research in combinatorics.

A code is said to be *t-error correcting* if, when no more than t errors have occurred in transmission of a code word, the transmitted code word can be recovered at the receiving end. A code containing two words at a distance one would not even be one-error correcting.

Theorem 7.3. A code is t-error correcting if its distance is at least $2t + 1$. Conversely, if a code is t-error correcting then its distance cannot be less than $2t + 1$.

Proof. Suppose we have a code of distance $2t + 1$ or more, and the word

x_1 was transmitted, and x_2 was received. If x_2 is not a code word the person at the receiving end will wonder what was sent. He will naturally choose the word in the code nearest to x_2. We shall show that if x_3 is any code word other than x_1, then $d(x_1, x_2) < d(x_3, x_2)$. Hence, by using the minimum distance rule he will correctly interpret x_2 as x_1.

Suppose $d(x_3, x_2) \leqslant t$. Since by hypothesis the number of errors is $\leqslant t$, then $d(x_1, x_2) \leqslant t$. Hence,

$$d(x_1, x_3) \leqslant d(x_1, x_2) + d(x_2, x_3) \leqslant t + t = 2t.$$

This is a contradiction since the distance between the code words x_1 and x_3 must be at least $2t + 1$. Hence x_1 is nearer to x_2 than x_3, and x_2 will be correctly interpreted as x_1.

Conversely, if a code is t-error correcting (using the minimum distance rule), then its distance cannot be less than $2t + 1$. Suppose two words x_1 and x_3 of the code are at a distance $\delta \leqslant 2t$. Then x_1 and x_3 disagree in δ positions and agree in all others. Let x_2 be the word obtained from x_3 by substituting in $[\delta/2]$ of the disagreeing positions, the corresponding letter of x_1. Suppose x_1 is transmitted and x_2 is received. The number of errors is $\delta - [\delta/2] \leqslant t$. Then $d(x_2, x_3) = [\delta/2] \leqslant \delta - [\delta/2] = d(x_1, x_2)$. Hence $d(x_2, x_3) \leqslant d(x_1, x_2)$. Thus x_2 will be wrongly interpreted as x_3 if the strict inequality holds, and if $d(x_2, x_3) = d(x_1, x_2)$ then the receiver will be left in doubt whether to interpret x_2 as x_3 or x_1. Hence, if the code has t-error correcting capability then its distance must be at least $2t + 1$.

There are various limits on the values of q (the size of the alphabet), n (the size of the word), t (the number of errors corrected), and N (the number of words in the code). It is clear that a packing problem is involved, because each code word we choose forces the exclusion of all "nearby" words from the code.

Theorem 7.4 The numbers q, n, t, and N are related by the inequality

$$Ns \leqslant q^n, \tag{7.5}$$

where

$$s = \binom{n}{0} + \binom{n}{1}(q-1) + \binom{n}{2}(q-1)^2 + \cdots + \binom{n}{t}(q-1)^t. \tag{7.6}$$

Proof. Let $S(x)$ denote the set of words at a distance t or less from a code word x. If x_1 and x_3 are any two code words then the set $S(x_1)$, is disjoint from $S(x_3)$, for if x_2 belongs to both $S(x_1)$ and $S(x_3)$ then

$$d(x_1, x_3) \leqslant d(x_1, x_2) + d(x_2, x_3) \leqslant t + t = 2t,$$

which is a contradiction since a t-error correcting code has distance at least $2t + 1$. But s is the number of words at a distance t or less from x. Hence $|S(x)| = s$. Thus there are at least Ns different words. Since the total number of different words is q^n, the result (7.5) follows.

A code in which (7.5) is an equality, so that $Ns = q^n$, is called *perfect*. We present here some examples of codes, based on structures we have studied.

Example 5. Any pair of orthogonal Latin squares of order n yields a 1-error correcting code with n^2 code words of length 4 over the alphabet $\{0, 1, 2, \ldots, n-1\}$. The code words are merely the 4-tuples of the form (i, j, a_{ij}, b_{ij}), $0 \leqslant i, j \leqslant n-1$, with $[a_{ij}] = A$ and $[b_{ij}] = B$ forming the two Latin squares. Suppose $\omega = (i, j, a_{ij}, b_{ij})$ and $\omega' = (i', j', a_{i'j'}, b_{i'j'})$ are two such words. If $i = i'$ and $j = j'$, clearly the two words are the same. Similarly, if $a_{ij} = a_{i'j'}$ and $b_{ij} = b_{i'j'}$, they must be the same word, since A and B are orthogonal. If $i = i'$ and $a_{ij} = a_{i'j'}$, then the words are the same because A is Latin. The other cases are all similar. Thus any two code words at distance 2 or less are the same, and we have a code of distance 3, which will correct one error.

Example 6. The Fano plane PG(2, 2) (Figure 7.10) yields a 1-error correcting code which is perfect. The points of the plane are numbered 1 to 7, and produce code words as follows.

The code words are of length 7 and the point i corresponds to the ith coordinate. Corresponding to any set of points, we take the ith coordinate to be 1 or 0 depending on whether the point i belongs or does not belong to the set.

Class 1. The 7-tuple corresponding to the null set

$$(0, 0, 0, 0, 0, 0, 0)$$

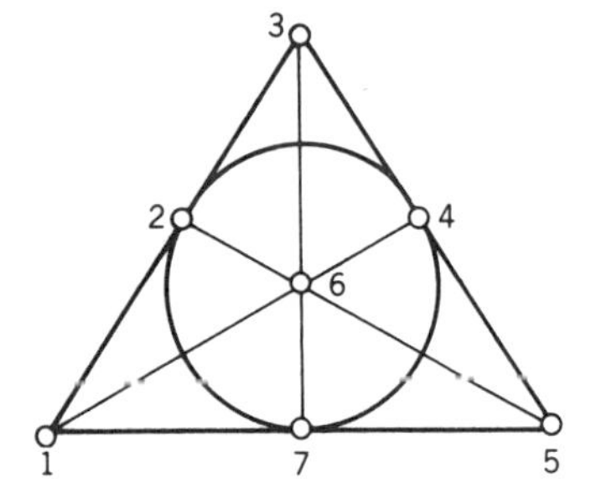

Figure 7.10

Class 2. The 7-tuples corresponding to points on a given line

$$(1, 1, 1, 0, 0, 0, 0)$$
$$(0, 0, 1, 1, 1, 0, 0)$$
$$(1, 0, 0, 0, 1, 0, 1)$$
$$(0, 1, 0, 0, 1, 1, 0)$$
$$(1, 0, 0, 1, 0, 1, 0)$$
$$(0, 0, 1, 0, 0, 1, 1)$$
$$(0, 1, 0, 1, 0, 0, 1).$$

Class 2′. The 7-tuples corresponding to points off a given line (complements of class 2)

$$(0, 0, 0, 1, 1, 1, 1)$$
$$(1, 1, 0, 0, 0, 1, 1)$$
$$(0, 1, 1, 1, 0, 1, 0)$$
$$(1, 0, 1, 1, 0, 0, 1)$$
$$(0, 1, 1, 0, 1, 0, 1)$$
$$(1, 1, 0, 1, 1, 0, 0)$$
$$(1, 0, 1, 0, 1, 1, 0)$$

Class 1′. The 7-tuple corresponding to the complete set (complement of class 1)

$$(1, 1, 1, 1, 1, 1, 1)$$

We leave it as an exercise for the reader to check that the sixteen 7-words obtained in this way form a code of distance 3, which is therefore 1-error correcting. We find s [using formula (7.6)] to be $\binom{7}{0} + \binom{7}{1}(2 - 1) = 1 + 7 = 8$. Thus $Ns = 16 \cdot 8 = 2^7 = q^n$, so the code is perfect.

EXERCISES

1. Show that the codes of Example 5 are not perfect for $n > 3$.
2. Give an argument that shows that the code of Example 6 has distance 3 by showing:

   (a) Any two code words corresponding to lines are at distance exactly 3.

   (b) Any two code words, one corresponding to a line and one the complement of a line, are at distance at least 3.

   (c) In a code over the alphabet $\{0, 1\}$, two words are at the same distance from each other as their "complements" are from each other.

(d) Finish the argument.

3. Can there exist a code with alphabet size $q = 4$, correcting $t = 2$ errors, in which each word is of size 5, and the number of words is 21? Give reasons.

4. A code is constructed according to the method of Example 5 using the orthogonal Latin squares.

	0	1	2
0	0	1	2
1	2	0	1
2	1	2	0

	0	1	2
0	0	1	2
1	1	2	0
2	2	0	1

(a) Write down the nine words of the code. Find the transmitted words corresponding to the following code words that are received:

$$(0, 1, 2, 2), \qquad (2, 2, 2, 0)$$

(b) What words, if any, are at distance 2 from each of the code words (1, 2, 1 0) and (2, 1, 2, 0)?

5. (a) In the code of Example 6, find the correct transmitted words corresponding to the following received words.

$$(0, 1, 1, 0, 1, 1, 0)$$
$$(1, 0, 0, 0, 1, 1, 0)$$
$$(0, 0, 0, 0, 1, 0, 0).$$

(b) If the word (1, 0, 1, 0, 0, 1, 0) is received, what would be your conclusion if you are using the minimum distance rule?

7.4. HAMMING ONE-ERROR CORRECTING BINARY CODES

In a *binary code* the alphabet consists of only two symbols 0 and 1, which may be regarded as elements of GF_2. Let Ω be the set of all binary r-vectors other than the null vector $(0, 0, \ldots, 0)$. Then there are $n = 2^r - 1$ vectors in Ω. Let H be the matrix whose columns are elements of Ω. Thus if $r = 3$, H is given by

$$H = \begin{bmatrix} 1 & 0 & 1 & 1 & 1 & 0 & 0 \\ 1 & 1 & 0 & 1 & 0 & 1 & 0 \\ 1 & 1 & 1 & 0 & 0 & 0 & 1 \end{bmatrix} \tag{7.7}$$

Let C be the *Hamming code* obtained by taking for code words all binary n-vectors $\mathbf{x}' = (x_1, x_2, \ldots, x_n)$ satisfying the equations

$$\mathbf{x}'H' = 0 \quad \text{or} \quad H\mathbf{x} = 0, \tag{7.8}$$

where H' is the transpose of H and $\mathbf{x}$ is the transpose of $\mathbf{x}'$.

The matrix H is called the *parity check matrix* of the code C. The rank of H is r. The rank cannot exceed r since H has r rows. On the other hand, the rank cannot be less than r since the last r column vectors are clearly independent. Hence there are $k = n - r = 2^r - 1 - r$ independent solutions of (7.8). Therefore, the code C contains 2^k words, namely all possible linear combinations of the k independent solutions.

We shall prove that the distance between any two code words of C is at least 3. Hence from Theorem 7.4, the code C is one-error correcting. We have only to show that two different code words cannot be at a distance 1 or 2. Let $\mathbf{u}'$ and $\mathbf{v}'$ be two distinct code words. Then from (7.8), we have $\mathbf{u}'H' = 0$, $\mathbf{v}'H' = 0$. Hence

$$(\mathbf{u}' - \mathbf{v}')H' = 0 \tag{7.9}$$

If $d(\mathbf{u}', \mathbf{v}') = 1$, let the ith coordinate of $\mathbf{u}' - \mathbf{v}'$ be 1 and let the other coordinates be 0. The left-hand side of (7.9) is $\mathbf{a}_i'$, where $\mathbf{a}_i'$ is the ith row of H'. This is a contradiction since the null vector is not an element of Ω and thus not a row of H'.

If $d(\mathbf{u}', \mathbf{v}') = 2$, let the ith and jth coordinates of $\mathbf{u}' - \mathbf{v}'$ be 1 and let the other coordinates be 0, $i \neq j$. Then from (7.9), $\mathbf{a}_i' = \mathbf{a}_j'$ where $\mathbf{a}_i'$ and $\mathbf{a}_j'$ are the ith and jth rows of H'. This is a contradiction since the rows of H' are all distinct. Hence $d(\mathbf{u}', \mathbf{v}') \geqslant 3$ and the code C is one-error correcting.

Note that the code C obtained here is perfect, since $q = 2$, $t = 1$, $n = 2^r - 1$, $N = 2^k$, where $k = n - r$. Hence

$$Ns = 2^k\{1 + \tbinom{n}{1}\} = 2^{k+r} = 2^n = q^n,$$

so that the equality holds in (7.5).

Example 7. Let $r = 3$. Then for the corresponding Hamming code, H is given by (7.7). Let

$$\mathbf{x}' = (x_1, x_2, x_3, x_4, x_5, x_6, x_7).$$

The equations (7.8) can now be written as

$$\begin{aligned} x_1 \quad + x_3 + x_4 &= x_5, \\ x_1 + x_2 \quad + x_4 &= x_6, \\ x_1 + x_2 + x_3 \quad &= x_7. \end{aligned}$$

Giving all possible values to x_1, x_2, x_3, x_4 we get the 16 words of the code. They are

$$\begin{array}{ll} (0,0,0,0,0,0,0) & (1,1,1,1,1,1,1) \\ (1,0,0,0,1,1,1) & (0,1,1,1,0,0,0) \\ (0,1,0,0,0,1,1) & (1,0,1,1,1,0,0) \\ (0,0,1,0,1,0,1) & (1,1,0,1,0,1,0) \\ (0,0,0,1,1,1,0) & (1,1,1,0,0,0,1) \\ (1,1,0,0,1,0,0) & (0,0,1,1,0,1,1) \\ (1,0,1,0,0,1,0) & (0,1,0,1,1,0,1) \\ (1,0,0,1,0,0,1) & (0,1,1,0,1,1,0) \end{array}$$

Suppose $\mathbf{x}' = (x_1, x_2, \ldots, x_n)$ is the word transmitted and $\mathbf{y}' = (y_1, y_2, \ldots, y_n)$ is the word received. The receiver will know only $\mathbf{y}'$. To use the minimum distance rule he or she will have to compare it with all the code words. If there is no error he or she will find one word in C, namely $\mathbf{x}'$ that matches $\mathbf{y}'$, and he or she will conclude that $\mathbf{x}'$ has been transmitted. Thus if in Example 7 the received word is (1, 1, 0, 0, 1, 0, 0), then since it is a code word, the conclusion is that this is the transmitted word. If, on the other hand, the received word is $\mathbf{y}' = (1, 0, 1, 0, 1, 0, 0)$, then the closest code word to it is $\mathbf{x}' = (1, 0, 1, 1, 1, 0, 0)$, which is at a distance 1 from $\mathbf{y}'$. The conclusion then will be that $\mathbf{x}'$ has been transmitted.

The number of comparisons needed for the minimum distance decoding rule is $N = 2^k = 2^{2^r - 1 - r}$. N increases rapidly with r; thus when $r = 4$, $N = 2^{11} = 2048$ and when $r = 5$, $N = 2^{26} = 77{,}108{,}864$. Thus the minimum distance rule becomes impractical when r is moderately large. We give here another decoding rule.

Let $\mathbf{x}'$ be the transmitted and $\mathbf{y}'$ the received words. Then the *error vector* is defined as the vector $\mathbf{e}' = (e_1, e_2, \ldots, e_i, \ldots, e_n)$, where $e_i = 1$ if there is an error in the ith coordinate and $e_i = 0$ if the ith coordinate has been correctly transmitted. Thus $w(\mathbf{e}')$ = number of errors in the transmission. Now $\mathbf{y}' = \mathbf{x}' + \mathbf{e}'$. Therefore, $\mathbf{y}'H' = \mathbf{x}'H' + \mathbf{e}'H'$. Since $\mathbf{x}'$ is a code word it follows from (7.8) that $\mathbf{x}'H' = 0$. Hence $\mathbf{y}'H' = \mathbf{e}'H'$. We define $\mathbf{y}'H'$ to be the *syndrome* of the received word $\mathbf{y}'$. If there is no error the syndrome is the null vector. If there is one error, say in the ith coordinate, then the syndrome is $\mathbf{h}_i'$, the ith row of H'. Hence our decoding rule is as follows.

If the syndrome $\mathbf{y}'H'$ is the null vector conclude that $\mathbf{y}'$ is the transmitted word. If the syndrome matches the ith row of H' conclude that there is an error in the ith coordinate and the transmitted word is $\mathbf{x}' = \mathbf{y}' - \mathbf{e}_i'$ where $\mathbf{e}_i' = (0, 0, \ldots, 1, \ldots, 0)$ [in $\mathbf{e}_i'$ the ith coordinate is 1, and the other coordinates are zero].

In Example 7 if the received word is $\mathbf{y}' = (1, 1, 0, 0, 1, 0, 0)$, then $\mathbf{y}'H' = (0, 0, 0, 0, 0, 0, 0)$ and the conclusion is that the transmitted word $\mathbf{x}' = \mathbf{y}'$. If the received word is $\mathbf{y}' = (1, 0, 1, 0, 1, 0, 0)$, then $\mathbf{y}'H' = (1, 1, 0)$. Since this matches the fourth row of H', then $\mathbf{e}' = (0, 0, 0, 1, 0, 0, 0)$. Then we conclude that the transmitted word is $\mathbf{x}' = \mathbf{y}' - \mathbf{e}' = (1, 0, 1, 1, 1, 0, 0)$.

Once the syndrome has been calculated the number of comparisons needed is $n = 2^r - 1$, which is 15 for $r = 4$ and 31 for $r = 5$.

EXERCISES

1. Show that there exists a matrix H_1 obtained by suitably permuting the columns of the matrix H given by (7.7), such that the code C_1 consisting of all 7-vectors $\mathbf{x}'$ (with elements from GF_2) satisfying $\mathbf{x}'H_1' = 0$ is identical with the code of Example 6 (this gives an alternative proof of the one-error correcting property of the code of Example 6).

2. The parity check matrix of a one-error correcting binary Hamming code is taken as:

$$\begin{bmatrix} 1\,0\,1\,1\,1\,1\,1\,1\,0\,0\,0\,1\,0\,0\,0 \\ 1\,1\,0\,1\,1\,1\,0\,0\,0\,1\,1\,0\,1\,0\,0 \\ 1\,1\,1\,0\,1\,0\,1\,0\,1\,0\,1\,0\,0\,1\,0 \\ 1\,1\,1\,1\,0\,0\,0\,1\,1\,1\,0\,0\,0\,0\,1 \end{bmatrix}$$

Answer the following questions.

(a) Supply the missing coordinates in the code word

$$(1, 1, 0, 1, 0, 1, 0, 1, 1, 1, 0, .\,, .\,, .\,, .).$$

(b) What is the conclusion regarding the transmitted word if the received word is:

(i) (1, 0, 1, 0, 1, 1, 0, 0, 1, 1, 1, 1, 1, 1, 1),
(ii) (1, 0, 1, 0, 1, 1, 0, 1, 0, 0, 1, 1, 1, 1, 1),
(iii) (1, 0, 1, 0, 1, 1, 0, 1, 1, 1, 1, 1, 1, 1, 1).

(c) How many different code words are there in the code with this parity check matrix?

7.5. REMARKS

In 1779 Euler proposed the problem:

> Arrange 36 officers, 6 from each of 6 regiments, of 6 different ranks, into a 6×6 square, so that each row and each file contains one officer of each rank and one officer of each regiment.

It is not hard to see that this is equivalent to finding a pair of orthogonal Latin squares of order 6. Since he could not do this, Euler conjectured that there is no pair of orthogonal squares of order 6, and in fact there is no pair of any order that is twice an odd number.

This conjecture was important, because it is not difficult to show (see Liu [1]) that if $p_1^{a_1}p_2^{a_2} \cdots p_t^{a_t}$ is the prime power decomposition of n, then there is a set of at least $k - 1$ orthogonal Latin squares of order n, if k is the minimum of the numbers $p_i^{a_i}$. The minimum of $p_i^{a_i}$ is at least 3, and therefore there is a pair of orthogonal squares of order n, unless n is twice an odd number.

In 1900 Tarry managed to list all Latin squares of order 6 and showed that Euler was correct that the officers problem had no solution. The general form of Euler's conjecture remained unsolved until 1960 when Bose, Shrikhande, and Parker surprised the mathematical world by showing that the rest of Euler's conjecture was wrong. In fact, orthogonal squares exist for orders 10, 14, 18, and so on. Thus orthogonal Latin squares exist of every order n except $n = 1$, 2, and 6. See Dênes and Keedwell [2] for more information on Latin squares.

Coding theory was invented, along with information theory, in 1948 by a group of mathematicians and scientists headed by Claude Shannon at Bell Telephone Laboratories. Since then the subject has grown by leaps and bounds because of its vital importance to communication and its mathematical elegance. Berlekamp [3], Peterson and Weldon [4], and Pless [5] contain useful introductions. The book of MacWilliams and Sloane [6] is encyclopedic but is above the level of our book. The topics of Chapter 6, 7, and 8 are all tied together nicely in Cameron and Van Lint [7].

[1] C. L. Liu, *Introduction to Combinatorial Mathematics*, McGraw-Hill, New York, 1968.

[2] J. Dênes and A. D. Keedwell, *Latin Squares and Their Applications*, English University Press, London, 1974.

[3] E. R. Berlekamp, *Algebraic Coding Theory*, McGraw-Hill, New York, 1968.

[4] W. W. Peterson and E. J. Weldon, Jr., *Error-Correcting Codes*, MIT Press, Cambridge, 1972.

[5] V. Pless, *Introduction to the Theory of Error-Correcting Codes*, Wiley, New York, 1982.

[6] F. J. MacWilliams and N. J. A. Sloane, *The Theory of Error-Correcting Codes*, North Holland, Amsterdam, 1977.

[7] P. J. Cameron and J. H. Van Lint, *Graph Theory, Coding, and Block Designs*, Cambridge University Press, Cambridge, 1975.

CHAPTER 8

Balanced Incomplete Block Designs

8.1. RELATIONS BETWEEN PARAMETERS

We shall begin this chapter with an example: Suppose we are required to arrange nine objects into sets of three such that every pair occurs in exactly one set.

With nine objects there are $\binom{9}{2} = 36$ pairs. Since each set of three objects yields three pairs, there must be exactly twelve sets. Also there are eight pairs containing any particular object θ. But if θ occurs in any set, then this set provides two pairs containing θ. Hence θ must occur in exactly four sets.

Thus, given nine objects, we have to find twelve sets such that each set contains three objects, each object occurs in four sets, and each pair of objects occurs together in exactly one set.

One way to obtain the sets is to identify the nine objects (denoted by 1, 2, 3, 4, 5, 6, 7, 8, 9) with the nine points of the finite geometry EG(2, 3). Then three objects are chosen to belong to the same set if the corresponding points lie on the same line. Since any two points are joined by exactly one line, every pair of objects will occur in exactly one set. We shall take the correspondence shown below:

Point of EG(2, 3)	Object
(0, 1)	1
(0, 2)	2
(1, 0)	3
(1, 1)	4
(1, 2)	5
(2, 0)	6

Point of EG(2, 3)	Object
(2, 1)	7
(2, 2)	8
(0, 0)	9

Using Table 6.2 and giving the lines of EG(2, 3), the required sets are shown in Table 8.1. Note that in Table 8.1 the three sets in the same row contain between them all nine objects.

Table 8.1

(9, 3, 6)	(1, 4, 7)	(2, 4, 8)
(9, 4, 8)	(1, 5, 6)	(2, 3, 7)
(9, 5, 7)	(1, 3, 8)	(2, 4, 6)
(9, 1, 2)	(3, 4, 5)	(6, 7, 8)

The general problem illustrated by the above example can be stated as follows:

There are v objects or *treatments*, which are to be arranged in b sets or *blocks* satisfying the following conditions:

(i) Each block contains k treatments.
(ii) Each treatment occurs in r blocks.
(iii) Every pair of treatments occurs together in λ blocks.

Such an arrangement, if it exists, is called a *balanced incomplete block* (*BIB*) *design*, with parameters (v, b, r, k, λ). Thus the sets obtained in our example form a BIB design with parameters (9, 12, 4, 3, 1). The names "treatment" and "block" come from the statistical theory of design of experiments.

The parameters v, b, r, k, λ are not all independent. In fact, we shall prove the following theorem.

Theorem 8.1. The parameters v, b, r, k, λ of a BIB design satisfy the relations

$$bk = vr, \qquad \lambda(v - 1) = r(k - 1). \tag{8.1}$$

Proof. Since each block contains k treatments, the total number of treatments occurring in the b blocks is bk. This number is also vr, since each of the v treatment occurs in r blocks. Hence $bk = vr$.

Again there are r different blocks in which a treatment θ appears. Each of these blocks contains $k - 1$ of the remaining $v - 1$ treatments. Hence the number of treatments other than θ occurring in these blocks is $r(k - 1)$. But this number is also $\lambda(v - 1)$ since each of the $v - 1$ treatments must occur in λ

of these blocks for property **(iii)** to be satisfied. Therefore $r(k-1) = \lambda(v-1)$.

Corollary 8.1. For a given v, k, λ the parameters r and b given by

$$r = \frac{\lambda(v-1)}{k-1}, \qquad b = \frac{\lambda v(v-1)}{k(k-1)} \tag{8.2}$$

must be integral.

For example, there cannot exist a BIB design with $v = 17$, $k = 3$, $\lambda = 1$ because if the design exists, $b = 136/3$, which is nonintegral.

Let θ and ϕ be two distinct treatments of a BIB design with parameters (v, b, r, k, λ). Then θ occurs in r blocks. The λ blocks in which θ and ϕ occur together must form a subset of these blocks. Hence

$$r \geqslant \lambda. \tag{8.3}$$

When the equality holds in (3.1), that is, when $r = \lambda$ it follows from (8.1) that

$$v = k, \qquad b = r = \lambda.$$

In this case, each of the v treatments occurs in each of the b blocks. This special case is known as a *randomized block design.* A BIB design that is not a randomized block design is called a *proper* BIB design. The necessary and sufficient condition for a BIB design to be proper is $r > \lambda$.

EXERCISES

1. Show that the following system of blocks is a BIB design and evaluate v, b, r, k, λ.

$$\begin{array}{cccc} (1, 2, 3), & (1, 4, 7), & (1, 5, 9), & (1, 6, 8), \\ (4, 5, 6), & (2, 5, 8), & (2, 6, 7), & (2, 4, 9), \\ (7, 8, 9), & (3, 6, 9), & (3, 4, 8), & (3, 5, 7). \end{array}$$

2. A design is formed by taking for treatments the sixteen numbers in the following scheme

$$\begin{array}{cccc} 1 & 2 & 3 & 4 \\ 5 & 6 & 7 & 8 \\ 9 & 10 & 11 & 12 \\ 13 & 14 & 15 & 16 \end{array}$$

in the following manner: The ith block consists of the six treatments (other than i) in the same row, or in the same column with i, $1 \leqslant i \leqslant 16$. Show that we get a BIB design and obtain all its parameters.

3. (a) Show that we get a BIB design by taking for blocks all k-sets of a v-set, $k < v$. Obtain its parameters (this design if called an unreduced BIB design).

(b) Write down the blocks of an unreduced BIB design with $v = 6$, $k = 4$.

4. Show that a BIB design with parameters (v, b, r, k, λ) can be augmented by repeating each block n times, $n > 1$, to create a new design. Calculate its parameters (such a design is called an n-plicate of the original design).

5. Can there exist a BIB design with the following parameters?

(a) $v = 25, k = 5, \lambda = 1$

(b) $v = 31, k = 4, \lambda = 1$.

8.2. INCIDENCE MATRIX OF A BIB DESIGN

Consider a BIB design with parameters (v, b, r, k, λ). Then the $v \times b$ matrix

$$N = (n_{ij})$$

is called the *incidence matrix* of the design if the element n_{ij} in the ith row and the jth column is 1 or 0, depending on whether the ith treatment occurs or does not occur in the jth block.

Example 1. Consider the BIB design with parameters (7, 7, 3, 3, 1) given below:

Block	Treatments in the Block
1	(1, 2, 4)
2	(2, 3, 5)
3	(3, 4, 6)
4	(4, 5, 7)
5	(5, 6, 1)
6	(6, 7, 2)
7	(7, 1, 3)

This design is obtained by identifying the points and lines of PG(2, 2)

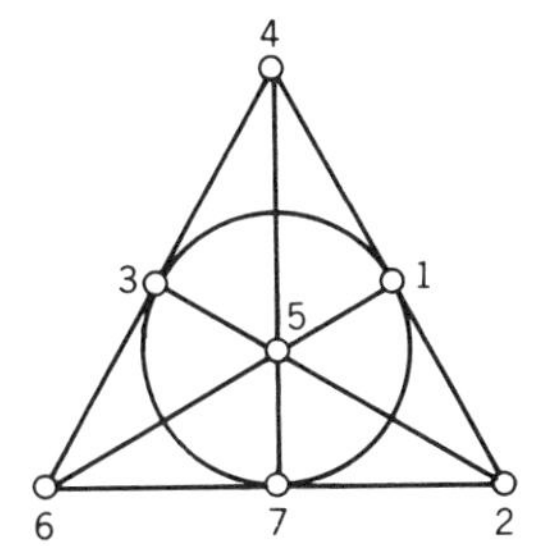

Figure 8.1

given in Table 6.3 with the treatments and blocks of the design. The correspondence between points and treatments is given below:

Treatment	Point
1	(1)
2	(1, 0)
3	(∞)
4	(0, 1)
5	(1, 1)
6	(0, 0)
7	(0)

Figure 7.10 of Chapter 7 may now be redrawn as shown in Figure 8.1.

The incidence matrix of the design is

$$
\begin{array}{c}
\begin{array}{ccccccc} 1 & 2 & 3 & 4 & 5 & 6 & 7 \end{array} \\
N = \begin{bmatrix}
1 & 0 & 0 & 0 & 1 & 0 & 1 \\
1 & 1 & 0 & 0 & 0 & 1 & 0 \\
0 & 1 & 1 & 0 & 0 & 0 & 1 \\
1 & 0 & 1 & 1 & 0 & 0 & 0 \\
0 & 1 & 0 & 1 & 1 & 0 & 0 \\
0 & 0 & 1 & 0 & 1 & 1 & 0 \\
0 & 0 & 0 & 1 & 0 & 1 & 1
\end{bmatrix}
\begin{array}{c} 1 \\ 2 \\ 3 \\ 4 \\ 5 \\ 6 \\ 7 \end{array}
\end{array}
$$

Theorem 8.2. If $N = (n_{ij})$ is the incidence matrix of a BIB design with parameters (v, b, r, k, λ), then

$$\text{(i)} \quad \sum_{j=1}^{b} n_{ij} = r,$$

$$\text{(ii)} \quad \sum_{i=1}^{v} n_{ij} = k.$$

Proof. Suppose i is fixed. Since there are r blocks that contain i, there are r values of j for which $n_{ij} = 1$, and $n_{ij} = 0$ for the other $b - r$ values of j. This proves (i).

Suppose j is fixed. Since there are k treatments contained in the block j, there are k values of i for which $n_{ij} = 1$, and $n_{ij} = 0$ for the other $v - k$ values of i. This proves (ii).

Corollary 8.2. In the incidence matrix N the sum of each row is r, and the sum of each column is k.

This is only a restatement of Theorem 8.2.

Theorem 8.3. If $N = (n_{ij})$ is the incidence matrix of a BIB design with parameters (v, b, r, k, λ), then

$$\text{(i)} \quad \sum_{j=1}^{b} n_{ij} n_{\alpha j} = \begin{cases} r & \text{if } i = \alpha, \\ \lambda & \text{if } i \neq \alpha, \end{cases}$$

$$\text{(ii)} \quad NN' = (r - \lambda) I_v + \lambda J_v,$$

where I_v is the unit matrix of order v, and J_v is the v by v matrix all of whose elements are unity.

$$\text{(iii)} \quad |NN'| = rk(r - \lambda)^{v-1}.$$

Proof of (*i*). If $i = \alpha$, then (*i*) becomes

$$\sum_{j=1}^{b} n_{ij}^2 = r.$$

But $n_{ij} = n_{ij}^2$ since $n_{ij} = 0$ or 1. Hence for the case $i = \alpha$ the required result follows from part (i) of Theorem 8.2.

Now suppose $i \neq \alpha$. Then $n_{ij} n_{\alpha j} = 1$ if and only if both n_{ij} and $n_{\alpha j}$ are unity, that is, both the treatments i and α occur in the block j. Now there are exactly λ values for j for which the pair of treatments i and α occur together in the block j. Hence $n_{ij}n_{\alpha j} = 1$ for exactly λ values of j for any fixed i and α, $i \neq \alpha$, and is zero for the other values of j. This proves (i) for the case $i \neq \alpha$.

Proof of (ii).

$$N = \begin{bmatrix} n_{11} & n_{12} & \dots & n_{1j} & \dots & n_{1b} \\ n_{21} & n_{22} & \dots & n_{2j} & \dots & n_{2b} \\ \dots & \dots & \dots & \dots & \dots & \dots \\ n_{v1} & n_{v2} & \dots & v_{vj} & \dots & n_{vb} \end{bmatrix}$$

From (i) it follows that

$$NN' = \begin{bmatrix} r & \lambda & \lambda & \dots & \lambda \\ \lambda & r & \lambda & \dots & \lambda \\ \lambda & \lambda & r & \dots & \lambda \\ \dots & \dots & \dots & \dots & \dots \\ \lambda & \lambda & \lambda & \dots & r \end{bmatrix} = \begin{bmatrix} r-\lambda & 0 & 0 & \dots & 0 \\ 0 & r-\lambda & 0 & \dots & 0 \\ 0 & 0 & r-\lambda & \dots & 0 \\ \dots & \dots & \dots & \dots & \dots \\ 0 & 0 & 0 & \dots & r-\lambda \end{bmatrix} +$$

$$\begin{bmatrix} \lambda & \lambda & \lambda & \dots & \lambda \\ \lambda & \lambda & \lambda & \dots & \lambda \\ \lambda & \lambda & \lambda & \dots & \lambda \\ \dots & \dots & \dots & \dots & \dots \\ \lambda & \lambda & \lambda & \dots & \lambda \end{bmatrix} = (r - \lambda)I_v + \lambda J_v.$$

Proof of (iii).

$$|NN'| = \begin{vmatrix} r & \lambda & \lambda & \dots & \lambda \\ \lambda & r & \lambda & \dots & \lambda \\ \lambda & \lambda & r & \dots & \lambda \\ \dots & \dots & \dots & \dots & \dots \\ \lambda & \lambda & \lambda & \dots & r \end{vmatrix} = \{r + \lambda(v - 1)\} \begin{vmatrix} 1 & 1 & 1 & \dots & 1 \\ \lambda & r & \lambda & \dots & \lambda \\ \lambda & \lambda & r & \dots & \lambda \\ \dots & \dots & \dots & \dots & \dots \\ \lambda & \lambda & \lambda & \dots & r \end{vmatrix}$$

by adding the last $v - 1$ rows to the first row and taking out the common factor $\{r + \lambda(v - 1)\}$ from the elements of the first row. Now multiplying the first row by λ and subtracting from the other rows, we have

$$|NN'| = \{r + \lambda(v - 1)\} \begin{vmatrix} 1 & 1 & 1 & \dots & 1 \\ 0 & r - \lambda & 0 & \dots & 0 \\ 0 & 0 & r - \lambda & \dots & 0 \\ \dots & \dots & \dots & \dots & \dots \\ 0 & 0 & 0 & \dots & r - \lambda \end{vmatrix}$$

$$= \{r + \lambda(v - 1)\}(r - \lambda)^{v-1}$$

$$= rk(r - \lambda)^{v-1}$$

from (8.1).

Corollary 8.3. For a proper BIB design $|NN'| \neq 0$, that is, NN' is non-singular.

Theorem 8.4. For a proper BIB design with parameters (v, b, r, k, λ),

$$b \geqslant v \quad \text{and} \quad r \geqslant k.$$

This result is often called Fisher's inequality, after R. A. Fisher.

Proof. Let N be the incidence matrix of the design. Suppose $b < v$. Let N_1 be the $v \times v$ matrix obtained from N by adding $v - b$ columns of zeros, that is,

$$N_1 = \begin{bmatrix} n_{11} & n_{12} & \dots & n_{1b} & 0 & \dots & 0 \\ n_{21} & n_{22} & \dots & n_{2b} & 0 & \dots & 0 \\ \dots & \dots & \dots & \dots & \dots & \dots & \dots \\ n_{v1} & n_{v2} & \dots & n_{vb} & 0 & \dots & 0 \end{bmatrix}.$$

Then

$$N_1 N_1' = NN'$$

$$|N_1 N_1'| = |NN'|$$

$$\neq 0 \qquad \text{from Corollary 8.3.}$$

But

$$|N_1 N_1'| = |N_1||N_1'|$$

$$= 0,$$

since N_1 has $v - b$ columns of zeros. We thus get a contradiction. Hence $b \geqslant v$. It follows from (2.1) that $r \geqslant k$.

Example 2. The BIB design with parameters $v = 16, b = 8, r = 3, k = 6, \lambda = 1$ cannot exist since Fisher's inequality is violated, even though the conditions of Theorem 8.1 are satisfied.

EXERCISES

1. Prove the converse of Theorem 8.3, part (ii), that is, if $N = (n_{ij})$ is a $v \times b$ matrix such that $n_{ij} = 0$ or 1 and $NN' = (r - \lambda)I_v + \lambda J_v$, then N is the incidence matrix of a BIB design.

2. Show that the matrix N given below is the incidence matrix of a BIB design. Obtain the parameters of this design.

$$N = \begin{bmatrix} 1 & 1 & 1 & 0 & 1 & 0 & 0 \\ 0 & 1 & 1 & 1 & 0 & 1 & 0 \\ 0 & 0 & 1 & 1 & 1 & 0 & 1 \\ 1 & 0 & 0 & 1 & 1 & 1 & 0 \\ 0 & 1 & 0 & 0 & 1 & 1 & 1 \\ 1 & 0 & 1 & 0 & 0 & 1 & 1 \\ 1 & 1 & 0 & 1 & 0 & 0 & 1 \end{bmatrix}.$$

3. Can there exist a BIB design with $v = 16$, $k = 10$, $\lambda = 3$?

4. A BIB design (v, b, r, k, λ) is said to be *resolvable* if the blocks can be divided into sets such that each treatment occurs exactly once in the blocks of any set. Show that the rank R of the incidence matrix of a resolvable design satisfies $R \leqslant b - r + 1$. Hence show that for a resolvable design Fisher's inequality can be improved to $b \geqslant v + r - 1$.

8.3. SYMMETRIC BIB DESIGNS

A BIB design is said to be *symmetric* if the number of blocks is equal to the number of treatments, that is,

$$v = b, \qquad r = k.$$

In this case the design is completely specified by the parameters k and λ and

$$v = b = \frac{k(k-1)}{\lambda} + 1. \tag{8.4}$$

Theorem 8.5. Any two blocks of a symmetric BIB design with parameters $v = b$, $r = k$, λ intersect in λ treatments.

Proof. Let N be the incidence matrix of the design. Then n is a symmetric $v \times v$ matrix. Since $r = k$ it follows from Theorem 8.2 that

$$NJ_v = J_vN = kJ_v. \tag{8.5}$$

Again from Theorem 8.3, part (ii) we have

$$\begin{aligned} NN'N &= \{(r-\lambda)I_v + \lambda J_v\}N \\ &= N\{(r-\lambda)I_v + \lambda J_v\} \\ &= NNN'. \end{aligned} \tag{8.6}$$

The theorem is trivially true for a symmetric BIB design, which is improper since $v = k$, and any two blocks intersect in k treatments. But $k = r = \lambda$. Hence we can assume that the design is proper and $r > \lambda$. From Corollary 8.3, $|NN'| > 0$. But for a symmetric design, N and N' are square matrices. Hence

$$|NN'| = |N| \cdot |N'| = |N|^2. \tag{8.7}$$

Hence $|N| \neq 0$. Therefore, N is nonsingular. We can, therefore, multiply both sides of (8.7) by N^{-1} and obtain

$$NN' = N'N. \tag{8.8}$$

If $j \neq \beta$, the element in the jth row and βth column of NN' is λ. But the element in the jth row and the βth column of $N'N$ is

$$\sum_{i=1}^{v} n_{ij}n_{i\beta} = \mu_{j\beta}, \tag{8.9}$$

where $\mu_{j\beta}$ is the number of treatments common to the blocks j and β, since $n_{ij}n_{i\beta}$ is unity if the treatment i occurs in both the blocks j and β and is zero otherwise. Hence $\mu_{j\beta} = \lambda$. This shows that any two blocks of a symmetric BIB design intersect in λ treatments.

Example 3. Given below are the blocks of the symmetric BIB design $v = b = 11$, $r = k = 5$, $\lambda = 2$. The method of obtaining these blocks is given later (Example 12).

1, 4, 5, 9, 3
2, *5*, 6, 10, *4*
3, 6, 7, 11, *5*
4, 7, 8, *1*, 6
5, 8, *9*, 2, 7
6, *9*, 10, *3*, 8
7, 10, 11, *4*, *9*,
8, 11, *1*, *5*, 10
9, *1*, 2, 6, 11
10, 2, *3*, 7, *1*
11, *3*, *4*, 8, 2

It is readily checked that any two blocks intersect in exactly two treatments. The treatments in which the first intersects the other blocks are shown in italics.

The conditions

$$bk = vr, \qquad \lambda(v - 1) = r(k - 1)$$

and Fisher's inequality

$$b \geqslant v$$

are necessary but not sufficient conditions for the existence of a BIB design with parameters v, b, r, k, λ. We can derive further necessary conditions in certain special cases, but no sufficient conditions are known for the general case. We shall now prove the following theorem.

Theorem 8.6. For a (proper) symmetric BIB design in which the number of treatments v is even, $r - \lambda$ must be a perfect square.

Proof. Let N be the incidence matrix of the design. From Theorem 8.3, part (iii),

$$\det(NN') = r^2(r - \lambda)^{v-1}.$$

However, for a symmetric design,

$$\det(NN') = (\det N)^2.$$

Since v is even, $r^2(r - \lambda)^{v-1}$ can be a perfect square only if $r - \lambda$ is a perfect square. This proves the theorem.

Example 4. The following symmetric BIB designs are impossible.

(i) $v = b = 22, r = k = 7, \lambda = 2.$
(ii) $v = b = 46, r = k = 10, \lambda = 2.$
(iii) $v = b = 92, r = k = 14, \lambda = 2.$
(iv) $v = b = 106, r = k = 15, \lambda = 2.$
(v) $v = b = 172, r = k = 19, \lambda = 2.$
(vi) $v = b = 34, r = k = 12, \lambda = 4.$
(vii) $v = b = 106, r = k = 21, \lambda = 4.$
(viii) $v = b = 52, r = k = 18, \lambda = 6.$
(ix) $v = b = 58, r = k = 19, \lambda = 6.$

For odd v, it can be shown, with a considerable amount of effort, that the diophantine equation $x^2 = (r - \lambda)y^2 + (-1)^{(v-1)/2}\lambda z^2$ has a solution in integers x, y, and z, not all zero, if a symmetric design exists. That criterion and Theorem 8.6 are generally known, together, as the Bruck–Chowla–Ryser criterion for existence of a symmetric BIB design.

EXERCISES

1. Given a BIB design D with parameters (v, b, r, k, λ) we obtain the *complementary design* D^* in the following manner. If B_i is a block of $D(1 \leqslant i \leqslant 6)$ then we take a block B_i^* of D^* such that a treatment belongs to B_i^* if and only if it does not belong to B_i. Show that D^* is a BIB design with parameters

$$v^* = v, \quad b^* = b, \quad r^* = b - r, \quad k^* = v - k, \quad \lambda^* = b - 2r + \lambda.$$

   Show that the complementary of a symmetric BIB design is itself symmetric with $r^* - \lambda^* = r - \lambda$.

2. Is it possible to have a symmetric BIB design with the following parameters?

$$v = b = 88, \quad r = k = 30, \quad \lambda = 10.$$

8.4. ORTHOGONAL SERIES DESIGNS

A BIB design with parameters

$$v = s^2, \quad b = s^2 + s, \quad r = s + 1, \quad k = s, \quad \lambda = 1, \tag{8.10}$$

where s is any integer, is said to belong to the *orthogonal series* 1 (OS1).

When s is a prime or a power of a prime we can at once construct the design given above by identifying the treatments with the points of EG(2, s) and the blocks with the lines of EG(2, s). This was illustrated in Section 8.1, where we obtained the BIB design (9, 12, 4, 3, 1) in this manner. No design belonging to the series OS1 is known when s is not a prime or a prime power, though the nonexistence has not been proved except for special values of s. The smallest value of s for which the question is still open is $s = 10$.

Instead of the geometry EG(2, s) it is more convenient to use a complete set of orthogonal Latin squares of order s, to obtain the orthogonal series design (8.10).

Example 5. We shall now illustrate the use of orthogonal Latin squares in constructing the orthogonal series 1 design with parameters

$$v = 16, \quad b = 20, \quad r = 5, \quad k = 4, \quad \lambda = 1.$$

We take three mutually orthogonal Latin squares $[L_1]$, $[L_2]$, $[L_3]$ of order 4, say the squares shown in Figure 7.7 and reproduced in Figure 8.2a,

$[L_1]$	$[L_2]$	$[L_3]$
0 1 2 3	0 1 2 3	0 1 2 3
1 0 3 2	2 3 0 1	3 2 1 0
2 3 0 1	3 2 1 0	1 0 3 2
3 2 1 0	1 0 3 2	2 3 0 1

Figure 8.2a

$[L_R]$	$[L_C]$
0 0 0 0	0 1 2 3
1 1 1 1	0 1 2 3
2 2 2 2	0 1 2 3
3 3 3 3	0 1 2 3

Figure 8.2b

and add to these the two squares $[L_R]$ and $[L_C]$ of Figure 8.2b, where the row i of L_R contains the symbol i in each cell and the column j of L_C contains the symbol j in each cell.

The sixteen treatments can be made to correspond to the cells of a 4×4 square as shown in Figure 8.3.

1	2	3	4
5	6	7	8
9	10	11	12
13	14	15	16

Figure 8.3

We now superpose the square of Figure 8.3 on any of the squares $[L_\alpha]$, $\alpha = R, C, 1, 2, 3$, and take those treatments in the same block which fall together with the same symbol of $[L_\alpha]$. The blocks obtained in this way are given below:

(1, 2, 3, 4), (5, 6, 7, 8), (9, 10, 11, 12), (13, 14, 15, 16),
(1, 5, 9, 13), (2, 6, 10, 14), (3, 7, 11, 15), (4, 8, 12, 16),
(1, 6, 11, 16), (2, 5, 12, 15), (3, 8, 9, 14), (4, 7, 10, 13),
(1, 7, 12, 14), (2, 8, 11, 13), (3, 5, 10, 16), (4, 6, 9, 15),
(1, 8, 10, 15), (2, 7, 9, 16), (3, 6, 12, 13), (4, 5, 11, 14).

The blocks obtained from each square are written in a separate row. Then each treatment appears exactly once in the blocks of a given row.

The process illustrated in the previous example is general. We start from $s - 1$ mutually orthogonal squares $[L_1], [L_2], \ldots, [L_{s-1}]$ and add to these the squares $[L_R]$ and $[L_C]$ such that the cell (i, j) of L_R contains the symbol i and the cell (i, j) of L_C contains the symbol j. The s^2 treatments can be made to correspond to the cells of an $s \times s$ square $[S]$. We now superpose the square $[S]$ on any of the squares $[L_\alpha]$, $\alpha = R, C, 1, 2, \ldots, s - 1$, and take those treatments in the same block which fall together with the same symbol of $[L_\alpha]$. It is obvious that we have

$$v = s^2, \quad b = s^2 + s, \quad r = s + 1, \quad k = s.$$

Thus, we have only to prove that $\lambda = 1$. Note that even though $[L_R]$ and $[L_C]$ are not Latin squares the squares $L_R, L_C, L_1, \ldots, L_{s-1}$ are orthogonal in the sense that if any one is superposed on the other, each symbol of the first occurs together with each symbol of the second exactly once.

Suppose θ and ϕ are any two treatments occurring in the cells (i_1, j_1) and

(i_2, j_2) of S. Then they would occur together in some block of our design if and only if there is a corresponding square $[L_t]$, $t \in \{R, C, 1, 2, \ldots, s-1\}$ such that the cells (i_1, j_1), (i_2, j_2) contain the same symbol, say u. There cannot be another square $[L_{t'}]$, $t' \in \{R, C, 1, 2, \ldots, s-1\}$, such that the cells (i_1, j_1) and (i_2, j_2) contain the same symbol, say u'. Otherwise, the symbol u of $[L_t]$ would occur with the symbol u' of $[L_{t'}]$ twice when $[L_t]$ and $[L_{t'}]$ are superposed. This, however, violates orthogonality. Hence, θ and ϕ cannot occur together in more than one block. But the number of pairs of treatments is $s^2(s^2-1)/2$. On the other hand, each of the $s(s+1)$ blocks of the design yields $s(s-1)/2$ pairs, that is, there are $s^2(s^2-1)/2$ pairs altogether. Hence, each pair of treatments occurs in exactly one block.

A BIB design with parameters

$$v = b = s^2 + s + 1, \quad r = k = s + 1, \quad \lambda = 1, \tag{8.11}$$

where s is any integer said to belong to the orthogonal series 2 (OS2).

When s is a prime or a prime power we can at once construct the design given above by identifying the treatment with the points of PG(2, s) and the blocks with the lines of PG(2, s). This was illustrated in Example 1 for the case $s = 2$, where the design (7, 7, 3, 3, 1) was obtained from PG(2, 2). But we can also use a complete set of orthogonal Latin squares to construct the design. We first construct the design (8.10). Then we take $s + 1$ new treatments, $t_R, t_C, t_1, t_2, \ldots, t_{s-1}$ corresponding to the squares $[L_R]$, $[L_C]$, $[L_1]$, $\ldots$, $[L_{s-1}]$. To each block of (8.10) obtained from $[L_\alpha]$ we add the treatment t_α, and then take a new block $(t_R, t_C, t_1, \ldots, t_{s-1})$. Now obviously there are $s^2 + s + 1$ blocks and treatments, each block contains $s + 1$ treatments, and each treatment occurs in $s + 1$ blocks. Also, each of the old pairs occurs exactly once. Since the blocks of (8.10) arising from $[L_\alpha]$ contain each treatment exactly once, each pair (θ, t_u) where θ is one of the old treatments occurs once. Also, any pair formed by the new treatments occurs in the new block. Thus every pair occurs exactly once. We thus get the design (8.11) from (8.10). Note that this process is exactly the same by which we obtained PG(2, s) from EG(2, s).

Example 6. We shall illustrate by obtaining the BIB design with parameters

$$v = b = 21, \quad r = k = 5, \quad \lambda = 1$$

from the design of Example 5. Let the new treatments t_R, t_C, t_1, t_2, t_3 be called 17, 18, 19, 20, and 21. Then the blocks of the required design are

(1, 2, 3, 4, 17), (5, 6, 7, 8, 17), (9, 10, 11, 12, 17), (13, 14, 15, 16, 17),
(1, 5, 9, 13, 18), (2, 6, 10, 14, 18), (3, 7, 11, 15, 18), (4, 8, 12, 16, 18),
(1, 6, 11, 16, 19), (2, 5, 12, 15, 19), (3, 8, 9, 14, 19), (4, 7, 10, 13, 19),
(1, 7, 12, 14, 20), (2, 8, 11, 13, 20), (3, 5, 10, 16, 20), (4, 6, 9, 15, 20),
(1, 8, 10, 15, 21), (2, 7, 9, 16, 21), (3, 6, 12, 13, 21), (4, 5, 11, 14, 21),

(17, 18, 19, 20, 21).

EXERCISES

1. Show that if s is a power of a prime we can construct a BIB design with parameters

(a) $v^* = b^* = s^2 + s + 1,\ r^* = k^* = s^2,\ \lambda^* = s^2 - s;$

(b) $v_1 = s^2,\ b_1 = s^2 + s,\ r_1 = s^2 - 1,\ k_1 = s^2 - s,\ \lambda_1 = s^2 - s - 1.$

(Use exercise 1, Section 8.3.)

2. Obtain BIB designs with parameters

$$v = 25, \quad b = 30, \quad r = 6, \quad k = 5, \quad \lambda = 1$$

and

$$v_0 = b_0 = 31, \quad r_0 = k_0 = 6, \quad \lambda_0 = 1$$

by using the orthogonal Latin squares of Example 2, Chapter 7.

3. A *balanced design* of index unity is a design with v_1 treatments, arranged into b_1 blocks not necessarily of the same size, such that any pair of treatments occurs together in the same block exactly once. Show that given a resolvable BIB design with parameters $v, b, r, k, \lambda = 1$, we obtain a balanced design of index unity with $v_1 = v + r$, $b_1 = b + 1$ in which there are b blocks of size $k + 1$, and one block of size r.

8.5. SYMMETRICALLY REPEATED DIFFERENCES

Consider a module M with v elements. Let each element of M correspond to a treatment. If there is a block $(a_1, a_2, \ldots, a_k)$ then the $k(k-1)$ elements $a_i - a_j$ are said to be the *differences* arising from the block. Of course, in particular the module M may be a ring or a field.

If there is a set of t blocks $B_1, B_2, \ldots, B_t$ $(t \geqslant 1)$ such that among the differences arising from them each nonzero element of M occurs λ times, then we say that differences are *symmetrically repeated* λ times.

Example 7. Let the treatments correspond to the elements of the field GF_7, the field of residue classes (mod 7). Consider the block (1, 2, 4). Then the differences arising from the block are

$$1 - 2 = 6, \quad 2 - 1 = 1, \quad 1 - 4 = 4, \quad 4 - 1 = 3, \quad 2 - 4 = 5, \quad 4 - 2 = 2.$$

Thus among the differences arising from the block each nonzero element of GF_7 occurs exactly once. Hence we say that the differences arising from the block are symmetrically repeated once.

Example 8. Let the treatments correspond to the field GF_{13}, the field of residue classes (mod 13). Then the differences arising from the two blocks (1, 3, 9) and (2, 6, 5) are 11, 2, 5, 8, 7, 6 and 9, 4, 10, 3, 1, 12. Thus each nonzero element of GF_7 occurs exactly once. Hence the differences arising from the set of blocks (1, 3, 9) and (2, 6, 5) are symmetrically repeated once.

Note that if B_1 and B_2 are two blocks such that the elements of B_2 are obtained by multiplying the elements of B_1 by $c \neq 0$, then if R is a field, the differences arising from B_2 are obtained by multiplying the differences arising from B_1 by c. Thus if a_i, a_j are any two elements of B_1 then the corresponding elements of B_2 are ca_i and ca_j. Our observation follows by noting that $ca_i - ca_j = c(a_i - a_j)$. In Example 8 the elements of the second block are obtained by multiplying the elements of the first block by 2. It can be checked that the differences arising from the second block are obtained by multiplying the elements of the first block by 2.

Given any block $B = (a_1, a_2, \ldots, a_k)$ with elements belonging to a ring M, the set of blocks $B_\theta = (a_1 + \theta, a_2 + \theta, \ldots, a_k + \theta)$, where θ takes all values of R, is said to be the set of blocks obtained by *developing B*.

Example 7 (continued). The set of blocks obtained by developing $B = (1, 2, 4)$ are

$$(1, 2, 4), \quad (2, 3, 5), \quad (3, 4, 6), \quad (4, 5, 7), \quad (5, 6, 1), \quad (6, 7, 2), \quad (7, 1, 3),$$

where instead of class 0 we have written class 7. This is permissible since (0) = (7).

Theorem 8.7. If in a BIB design the treatments correspond to the elements of a module M with v elements, and the differences arising from a set of t initial blocks $B_1, B_2, \ldots, B_t$ each containing k elements are symmetrically repeated λ times, then by developing those blocks we obtain a BIB design with parameters

$$v, \quad b = vt, \quad r = kt, \quad k, \quad \lambda.$$

Before proving the theorem we shall illustrate by an example.

Example 8 (continued). The blocks obtained by developing the two blocks (1, 3, 9) and (2, 6, 5) can be written as

$$\begin{array}{rr}
(1,\ 3,\ 9), & (2,\ 6,\ 5),\\
(2,\ 4,\ 10), & (3,\ 7,\ 6),\\
(3,\ 5,\ 11), & (4,\ 8,\ 7),\\
(4,\ 6,\ 12), & (5,\ 9,\ 8),\\
(5,\ 7,\ 13), & (6,\ 10,\ 9),\\
(6,\ 8,\ 1), & (7,\ 11,\ 10),\\
(7,\ 9,\ 2), & (8,\ 12,\ 11),\\
(8,\ 10,\ 3), & (9,\ 13,\ 12),\\
(9,\ 11,\ 4), & (10,\ 1,\ 13),\\
(10,\ 12,\ 5), & (11,\ 2,\ 1),\\
(11,\ 13,\ 6), & (12,\ 3,\ 2),\\
(12,\ 1,\ 7), & (13,\ 4,\ 3),\\
(13,\ 2,\ 8), & (1,\ 5,\ 4),
\end{array}$$

where instead of 0 we have written 13.

It is clear that in this design $v = 13$, $b = 26$, $k = 3$. Each treatment occurs exactly once in any column. Hence $r = 6$. We claim that each pair occurs exactly once. Corresponding to any pair a_i, a_j we have two differences $a_i - a_j$, $a_j - a_i$ whose sum is zero. Thus the differences corresponding to the pair 4, 7 are 10 and 3. Now since in the initial blocks the differences are symmetrically repeated there is exactly one pair giving rise to the differences 10 and 3. This pair is 2, 5. Now when we develop the blocks the corresponding differences remain the same. Thus the required pair 4, 7 must occur in the block $B_{2,\theta}$ obtained by developing B_2, where θ is given by $4 = 2 + \theta$, $7 = 5 + \theta$, that is, $\theta = 2$. Thus, 4, 7 occurs in the block (4, 8, 7). Thus $\lambda = 1$. We have therefore obtained the BIB design

$$v = 13, \quad b = 26, \quad r = 6, \quad k = 3, \quad \lambda = 1.$$

Proof. It is clear that the above argument is general. We arrange the t initial blocks in a row and write the developed blocks under them as shown below

$$\begin{array}{cccc}
B_1 & B_2 & \cdots & B_t\\
\cdots & \cdots & \cdots & \cdots\\
B_{1,\theta} & B_{2,\theta} & \cdots & B_{t,\theta}\\
\cdots & \cdots & \cdots & \cdots
\end{array}$$

where the elements of $B_{i,\theta}$ are obtained by adding θ to the elements of B_i. In the design so obtained it is clear that the first four parameters are v, $b = vt$, $r = kt$, k. So we have only to prove the constancy of λ. Consider the pair of treatments c, d. The differences arising from this pair are $c - d$ and $d - c$. Now there are exactly λ pairs in the initial blocks which give rise to the same differences. Let a_i, a_j be one such pair occurring, say, in the block B_u. Then $a_i - a_j = c - d$, $a_j - a_i = d - c$. If $a_i + \theta = c$, then $a_j + \theta = d$. Hence the pair c, d occurs in the block $B_{u,\theta}$, where $\theta = c - a_i = d - a_j$. Thus corresponding to each of the pairs in the initial blocks which give rise to the differences $c - d$ and $d - c$, there will be one block of the design where the pair c, d occurs. By developing the initial blocks $B_1, B_2, \ldots, B_t$ we shall therefore get a BIB design with the parameters

$$v, \quad b = vt, \quad r = kt, \quad k, \quad \lambda.$$

The conditions (8.1) are automatically satisfied. Clearly $bk = vr$. Also, each initial block gives $k(k - 1)$ differences. There are $v - 1$ nonzero elements, each appearing λ times among the differences $k(k - 1)t = (v - 1)\lambda$, that is $r(k - 1) = \lambda(v - 1)$.

EXERCISES

1. If the set of treatments belong to the ring of residue classes (mod 21), show that the differences arising from the block

$$3, 6, 12, 7, 14$$

are symmetrically repeated once. Hence obtain the BIB design with parameters (21, 21, 5, 5, 1).

2. Show that a solution of the BIB design with parameters (16, 80, 15, 3, 2) is obtained by developing the initial blocks

$$(0, 1, 3), \quad (0, 3, 8), \quad (0, 2, 12), \quad (0, 1, 7), \quad (0, 4, 9)$$

(mod 16).

3. Show that a solution of the BIB design with parameters (13, 39, 15, 5, 5) is obtained by developing the initial blocks

$$(0, 1, 2, 4, 8), \quad (0, 3, 6, 12, 11), \quad (0, 4, 8, 3, 6)$$

(mod 13).

8.6. STEINER TRIPLES

A *Steiner triple system* is an arrangement of v objects or treatments in triplets, such that every pair of objects appears in exactly one triplet. It is thus a BIB design with $k = 3$, $\lambda = 1$. From (8.1) we have

$$3b = vr, \qquad v - 1 = 2r$$

$$\therefore v = 2r + 1, \qquad b = \frac{r(2r + 1)}{3}.$$

Hence r must be of the form $3t$ or $3t + 1$. We therefore get the following two series of Steiner triple systems

$$v = 6t + 1, \quad b = t(6t + 1), \quad r = 3t, \quad k = 3, \quad \lambda = 1, \tag{8.12}$$

$$v = 6t + 3, \quad b = (3t + 1)(2t + 1), \quad r = 3t + 1, \quad k = 3, \quad \lambda = 1. \tag{8.13}$$

We shall now consider the series (8.12) and obtain a solution in the special case when $v = 6t + 1 = p^n$, where p is a prime. There exists a field GF_{p^n} with p^n elements. The treatments may be identified with the elements of this field. Let x be a primitive element of this field. Then

$$x^{v-1} = 1 \quad \text{or} \quad x^{6t} = 1 \tag{8.14}$$

and all the nonzero elements of the field are given by

$$x^0 = 1, \quad x, \quad x^2, \ldots, x^{6t-1}$$

Now from (8.14)

$$(x^{3t} + 1)(x^{3t} - 1) = 0.$$

Since x is a primitive element of the field, $x^{3t} \neq 1$. Hence

$$x^{3t} + 1 = 0 \quad \text{or} \quad x^{3t} = -1 \tag{8.15}$$

$$\therefore (x^t + 1)(x^{2t} - x^t + 1) = 0.$$

Since $x^t + 1 \neq 0$ (otherwise $x^{2t} = 1$, contradicting the fact that x is a primitive element of the field), we have

$$x^{2t} + 1 = x^t. \tag{8.16}$$

Let us start with the set of t initial blocks

$$(x^i, x^{2t+i}, x^{4t+i}), \qquad i = 0, 1, \ldots, t-1. \tag{8.17}$$

If we set

$$x^{2t} - 1 = x^q, \tag{8.18}$$

then the six differences arising from the typical initial block are

$$\pm x^i(x^{2t} - 1), \quad \pm x^i(x^{4t} - x^{2t}).$$

From (8.15), (8.16), and (8.18), we have

$$\begin{aligned}
&x^i(x^{2t} - 1) = x^{q+i}, \quad x^i(x^{4t} - 1) = x^{q+t+i}, \quad x^i(x^{4t} - x^{2t}) = x^{q+2t+i},\\
&-x^i(x^{2t} - 1) = x^{q+3t+i}, \quad -x^i(x^{4t} - 1) = x^{q+4t+i},\\
&-x^i(x^{4t} - x^{2t}) = x^{q+5t+i}.
\end{aligned}$$

Remembering (8.14) we see that, among the differences arising from the initial blocks (8.17), every nonzero element of GF_{p^n} is repeated exactly once. Hence from Theorem 8.7 we have

Theorem 8.8. The initial blocks (8.17) provide a solution of the BIB design of series (8.12) with parameters

$$v = 6t + 1, \quad b = t(6t + 1), \quad r = 3t, \quad k = 3, \quad \lambda = 1$$

if $6t + 1 = p^n$, where p is a prime.

Example 9. Let $t = 1$. Then $v = 7, b = 7, r = 3, k = 3, \lambda = 1$. We have to consider the field GF_7, a primitive element of which is 3. The initial block is $(3^0, 3^2, 3^4)$ or $(1, 2, 4)$. The complete solution is given by the set of blocks obtained by developing the initial blocks

$$(1, 2, 4), \quad (2, 3, 5), \quad (3, 4, 6), \quad (4, 5, 7), \quad (5, 6, 1), \quad (6, 7, 2), \quad (7, 1, 3).$$

This design also belongs to the orthogonal series OS2, with parameters given by (8.11), where $s = 1$. It is the same as the design of Example 1.

Example 10. Let $t = 2$. Then $v = 13$, $b = 26$, $r = 6$, $k = 3$, $\lambda = 1$. A primitive element of GF_{13} is 2. The initial blocks are

$$(2^0, 2^4, 2^8), \quad (2, 2^5, 2^9),$$

or

$$(1,\ 3,\ 9\ ),\quad (2, 6,\ 5\ ).$$

The complete solution is obtained by developing these initial blocks. It is the same as the design of Example 8.

Example 11. Let $t = 4$. Then $v = 25$, $b = 100$, $r = 12$, $k = 3$, $\lambda = 1$. This case is of some interest because v is a power of a prime without itself being a prime. The 25 treatments can be taken as the elements of GF_{5^2}. If x is a primitive element of this field, the initial blocks are

$$(x^0, x^8, x^{16}),\quad (x, x^9, x^{17}),\quad (x^2, x^{10}, x^{18}),\quad (x^3, x^{11}, x^{19}).$$

A minimum function for GF_{5^2} is $x^2 + 2x + 3$. Using this, the elements of the field can be represented as polynomials $ax + b$, where a and b are themselves residue classes (mod 5). (See Exercise 5, Section 5.5.) The initial blocks can therefore be written

$$(1, 4x + 1, x + 3), (x, 3x + 3, x + 2), (3x + 2, 2x + 1, 2),$$
$$(x + 1, 2x + 4, 2x).$$

The complete solution can be obtained by adding to these the polynomials $ax + b(a, b = 0, 1, 2, 3, 4)$ and reducing the coefficients (mod 5).

Here we have given a solution of the series (8.12) for the special case $6t + 1 = p^n$. However, it is known that a solution can be obtained in every case for both the series (8.12) and (8.13).

EXERCISES

1. Obtain the initial blocks from which a solution of the Steiner triple system with
$$v = 37,\quad b = 222,\quad r = 18,\quad k = 3,\quad \lambda = 1$$
can be obtained by developing the initial blocks (mod 37).

2. Obtain the initial blocks from which a solution of the Steiner triple system with
$$v = 31,\quad b = 155,\quad r = 15,\quad k = 3,\quad \lambda = 1$$
can be obtained by developing the initial blocks (mod 31).

3. Show that a solution of the Steiner triple system
$$v = 25,\quad b = 100,\quad r = 12,\quad k = 3,\quad \lambda = 1$$

can be obtained by developing the initial blocks

$$(0, 1, 3), \quad (0, 4, 13), \quad (0, 5, 11), \quad (0, 7, 17)$$

(mod 25). Note that this solution is different from the one given in Example 11.

8.7. SYMMETRIC BIB DESIGNS WITH $r = (v - 1)/2$

By definition $v = b$, $r = k$. Hence if $r = (v - 1)/2$, from (8.1),

$$\lambda(v - 1) = r(r - 1)$$
$$= \frac{v - 1}{2} \frac{v - 3}{2}$$
$$\therefore v = 4\lambda + 3.$$

Taking $\lambda = t - 1$, we see that the designs under construction belong to the series with parameters

$$v = b = 4t - 1, \quad r = k = 2t - 1, \quad \lambda = t - 1. \tag{8.19}$$

We shall first show that a solution is possible when $4t - 1$ is a prime power. Let $4t - 1 = p^n$. Let the treatments correspond to the elements of the field GF_{p^n}, and let the nonzero elements be

$$x^0 = 1, \quad x, \quad x^2, \ldots, x^{4t-3},$$

where x is a primitive element. We see as in (8.15) that

$$x^{2t-1} = -1. \tag{8.20}$$

Now consider the initial block

$$(x^0, x^2, \ldots, x^{4t-4}). \tag{8.21}$$

The differences arising from this initial block can be exhibited as

$$\pm(x^2 - x^0), \pm(x^4 - x^2), \ldots\ldots \pm(x^0 - x^{4t-4}) = \pm(x^{4t-2} - x^{4t-4}),$$
$$\pm(x^4 - x^0), \pm(x^6 - x^2), \ldots\ldots \pm(x^2 - x^{4t-4}) = \pm(x^{4t} - x^{4t-4}),$$
$$\cdots \qquad \cdots \qquad \cdots\cdots \qquad \cdots \qquad \cdots$$

$$\pm(x^{2t-2} - x^0),\ \pm(x^{2t} - x^2), \ldots\ldots \pm(x^{2t-4} - x^{4t-4}) = \pm(x^{6t-6} - x^{4t-4}).$$

If we set $x^{2i} - x^0 = x^{q_i}$, then remembering (8.20) the differences occurring in the ith row of the scheme above are

$$x^{q_i},\ x^{q_i+2}, \ldots,\ x^{q_i+4t-2},\ x^{q_i+2t-1},\ x^{q_i+2t+1}, \ldots,\ x^{q_i+6t-3}. \qquad (8.22)$$

But all the nonzero elements of GF_{p^n} occur exactly once in (8.22). Hence among the differences arising from the initial block (8.24), each difference arises exactly $t - 1$ times. It follows from Theorem 8.7 that a solution of the design is obtained by developing the initial block (8.21). Hence we have the following theorem.

Theorem 8.9. The initial block (8.21) provides a solution of the BIB designs of the series (8.19) with parameters

$$v = b = 4t - 1, \quad r = k = 2t - 1, \quad \lambda = t - 1$$

if $4t - 1$ is a prime power.

Example 12. Let $t = 3$. Then $v = b = 11$, $r = k = 5$, $\lambda = 2$. One primitive element of GF_{11} is 2. Hence the solution is obtained by developing (mod 11) the initial block

$$(1, 4, 5, 9, 3).$$

This is the same as the design of Example 3.

Example 13. Let $t = 5$. Then $v = b = 19$, $r = k = 9$, $\lambda = 4$. A primitive element of GF_{19} is 2. Hence the solution is obtained by developing (mod 19) the initial block

$$(1, 4, 16, 7, 9, 17, 11, 6, 5).$$

Example 14. Let $t = 17$. Then $v = b = 27$, $r = k = 13$, $\lambda = 6$. If x is a primitive root of GF_{3^3}, the solution is given by the initial block

$$(x^0, x^2, x^4, x^6, x^8, x^{10}, x^{12}, x^{14}, x^{16}, x^{18}, x^{20}, x^{22}, x^{24}).$$

Using the minimum function $x^3 + 2x + 1$, this can be written as

$$(1, x^2, x^2 + 2x, x^2 + x + 1, 2x^2 + 2, x^2 + x, x^2 + 2, 2x, 2x + 1, x^2 + 2x + 1, 2x^2 + x + 1, 2x + 2, 2x^2 + 2x + 1).$$

The complete solution is obtained by the addition of the polynomials $ax^2 + bx + c(a, b, c = 0, 1, 2)$ and reducing the coefficients (mod 3).

EXERCISES

1. Obtain a solution of the symmetric BIB design

$$v = b = 23, \quad r = k = 11, \quad \lambda = 5.$$

2. Obtain the initial block from which a solution of the symmetric BIB design with parameters

$$v = b = 31, \quad r = k = 15, \quad \lambda = 7$$

can be obtained by developing (mod 31).

3. If x is a primitive element of GF_{2^4} show that a solution of the symmetric BIB design with parameters

$$v = b = 16, \quad r = k = 6, \quad \lambda = 2$$

can be obtained from the initial block

$$(0, x^0, x^3, x^6, x^9, x^{12})$$

by successively adding different elements of GF_{2^4}.

8.8. THE RESIDUAL AND DERIVED OF A SYMMETRIC BIB DESIGN

Consider a symmetric BIB design D with parameters

$$v = b, \quad r = k, \quad \lambda.$$

Choose any block of D as initial block and delete from D the initial block and the treatments contained in the initial block.

We shall prove that the remaining blocks and treatments form a BIB design, which is called the *residual* of the original design. Thus, consider the BIB design $v = b = 11$, $r = k = 5$, $\lambda = 2$ of Example 3. We take the first block to be the initial block, so we have to delete the treatments in italics from the other blocks. Thus the residual design is

(2, 6, 10),
(6, 7, 11),

(7, 8, 6),
(8, 2, 7),
(6, 10, 8),
(7, 10, 11),
(8, 11, 10),
(2, 6, 11),
(10, 2, 7),
(11, 8, 2).

It is readily checked that this is a BIB design with parameters (6, 10, 5, 3, 2). Consider now the general case. Since one block and k treatments have been deleted, the number of treatments and blocks remaining is given by

$$v^* = v - k, \qquad b^* = b - 1.$$

From Theorem 8.5 any two blocks have exactly λ treatments in common. Hence from each block, λ treatments are deleted. Thus after deletion each block contains

$$k^* = k - \lambda$$

treatments. Finally we note that treatments and pairs of treatments not occurring in the initial block remain undisturbed. Hence

$$r^* = r, \qquad \lambda^* = \lambda.$$

The process by which the new design is obtained is called the *process of block section*. Hence we have the following theorem.

Theorem 8.10. From a symmetric BIB design D with parameters $v = b$, $r = k$, λ we can obtain by the process of block section a new design D^* (the residual of D) whose parameters are

$$v^* = v - k, \quad b^* = b - 1, \quad r^* = r, \quad k^* = k - \lambda, \quad \lambda^* = \lambda.$$

Again starting from the symmetric BIB design D with parameters $v = b$, $r = k$, λ we can delete an initial block and retain from the other blocks only the treatments contained in the initial block. Then the number of retained treatments and blocks is given by

$$v' = k, \qquad b' = b - 1.$$

Also, since any two blocks of the original have exactly λ treatments in common, we have in the new design

$$k' = \lambda.$$

Since each of the retained treatments is contained once in the initial block, and also each pair of treatments occurs once in the initial block deleted, we have in the new design

$$r' = r - 1, \qquad \lambda' = \lambda - 1.$$

The process by which the new design has been obtained is called the *process of block intersection*, and the new design D' is called the *derived* of the original design. Hence we have the following theorem.

Theorem 8.11. From a symmetric BIB design D with parameters $v = b$, $r = k$, λ we can obtain by the process of block intersection a new BIB design D' (the derived of D) whose parameters are

$$v' = k, \quad b' = b - 1, \quad k' = \lambda, \quad r' = r - 1, \quad \lambda' = \lambda - 1.$$

Applying the process of block section to the series of symmetric BIB designs with parameters given by (8.19), we get designs of the series

$$v = 2t, \quad b = 4t - 2, \quad r = 2t - 1, \quad k = t, \quad \lambda = t - 1. \tag{8.23}$$

Hence from Theorems 8.9 and 8.10 we have the following theorem.

Theorem 8.12. A solution of the BIB design with parameters given by (8.23) can be obtained if $4t - 1$ is a prime power.

Example 15. Taking $t = 3$, 5, and 7 in (8.23) we have

(i) $v = 6, b = 10, r = 5, k = 3, \lambda = 2.$
(ii) $v = 10, b = 18, r = 9, k = 5, \lambda = 4.$
(iii) $v = 14, b = 26, r = 13, k = 7, \lambda = 6.$

The solutions of the designs with parameters given by **(i)**, **(ii)**, and **(iii)** can be obtained by block section from the symmetric BIB designs of Examples 12, 13, and 14.

Applying this process of block intersection to the series (8.19) of symmetric BIB designs we get the designs of the series with parameters

$$v = 2t - 1, \quad b = 4t - 2, \quad r = 2t - 2, \quad k = t - 1, \quad \lambda = t - 2 \tag{8.24}$$

Hence from Theorems 8.9 and 8.11 we have the following theorem.

Theorem 8.13. A solution of the BIB design with parameters given by (8.23) can be obtained if $4t - 1$ is a prime power.

Example 16. Taking $t = 5$ and 7 in (8.24) we have

(i) $v = 9, b = 18, r = 8, k = 4, \lambda = 3.$
(ii) $v = 13, b = 26, r = 12, k = 6, \lambda = 5.$

The solutions of the designs with parameters given by **(i)** and **(ii)** can be obtained by block intersection from the symmetric BIB designs of Examples 13 and 14.

EXERCISES

1. If D_1 and D_2 are complementary symmetric BIB designs, show that the residual of D_1 and the derived of D_2 are also complementary designs.
2. Obtain a solution of the BIB design with parameters
$$v = 12, \quad b = 22, \quad r = 11, \quad k = 6, \quad \lambda = 5.$$
(Use Exercise 1, Section 8.7.)
3. Obtain a solution of the BIB design with parameters
$$v = 15, \quad b = 30, \quad r = 14, \quad k = 7, \quad \lambda = 6.$$
(Use Exercise 2, Section 8.7.)

8.9. REMARKS

Steiner triple systems far predate the more general structures discussed in this chapter. Steiner proposed the problem of the existence of such systems in 1853, not knowing that Kirkman had solved the problem in 1847. In fact, it was originally raised by W. S. B. Woolhouse in a periodical called *The Ladies' and Gentlemen's Diary* in 1844.

The principles of statistical designs were formulated in 1925 by R. A. Fisher, and the use of BIB designs was introduced by Yates in 1936. In

1939 Bose published a paper that set in motion the modern study of such structures, based on Galois fields. Since then the theory has developed extensively, driven both by the need of statisticians for designs and by the appreciation of mathematicians for the beauty of the subject.

Further information about the actual application of designs in statistical experiments can be found in Cox [1]. For a proof of the result of Bruck, Chowla, and Ryser stated at the end of Section 8.3 see Ryser [2]. A somewhat different approach to designs can be found in Cameron and Van Lint [3].

[1] D. R. Cox, *Planning of Experiments*, Wiley, New York, 1958.

[2] H. J. Ryser, *Combinatorial Mathematics*, MAA, New York, 1963, pp. 111–113.

[3] P. J. Cameron and J. H. Van Lint, *Graph Theory, Coding and Block Designs*, Cambridge University Press, Cambridge, 1975.

CHAPTER 9

Problems of Choice

9.1. INTRODUCTION

In this chapter we shall consider some theorems that guarantee the existence of certain choices under appropriate hypotheses. We start with two examples.

Example 1. Show that among any six people, either there are three people who are mutual friends or there are three who do not know each other. Note that any person A must either know three other people or be a stranger to three, since there are five people besides A in the group. Let us suppose A knows B, C, and D, since the other case is similar. If any two of B, C, and D know each other, we have, with A, a group of three mutual friends. Otherwise B, C, and D form a group of three people who do not know each other.

Example 2. Two busy psychiatrists, Dr. Head and Dr. Shrinker, have hour-long appointments five days a week. Dr. Head works from 9 to 3 and Dr. Shrinker from 9 to 2. Because they never cure anyone, they are becoming bored with their schedules, which have not varied for two years. They each decide to change everyone's appointment to the same hour on an adjacent day or an adjacent hour on the same day. Can they each do that for all of their patients?

Dr. Head can. One possible method is just to switch the first two appointments each day, the next two, and the last two. For Dr. Shrinker, however, such a scheme does not work. In fact, no change is possible that satisfies the given conditions. Perhaps the easiest way to see that is to draw the schedule as a rectangle, broken up like a chess board, as in Figure 9.1. Note that the schedule contains 13 black hours and 12 white

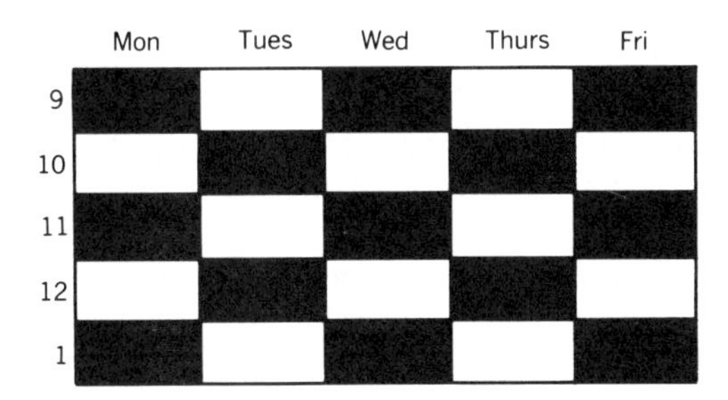

Figure 9.1

hours. Because the conditions given force each black appointment to be rescheduled into a white hour, and vice versa, no such rescheduling is possible for Dr. Shrinker.

In this chapter we will consider some general theorems that allow us to answer questions of the type illustrated by the examples. We shall also give some applications.

EXERCISES

1. Show that in any group of people, at least two members of the group have the same number of friends in the group.
2. Express the result of exercise 1 in graph-theoretic terminology.
3. Express the result of Example 1 in terms of triangles in a graph with six vertices.
4. For Example 2, is there a switching method for Dr. Head which changes everyone's appointment to the same hour on an adjacent day?
5. Can Dr. Shrinker's patients in Example 2 be rescheduled as required if exactly one patient is allowed to keep the same appointment slot? Does it make any difference which appointment slot is left unchanged?

9.2. THE PIGEONHOLE PRINCIPLE

A mathematical proof technique called the pigeonhole principle may be familiar to you, although you probably have not heard it explicitly stated. In its simplest form it states that if three pigeons nest in two holes, two pigeons must share a hole. For example, any group of three or more people contains two people of the same sex. This can be generalized in various ways. Increasing the number of holes, we note that if there are more pigeons than holes, two pigeons must share a hole. More exactly, if $|D| > |R|$ there is no one-to-one function from D to R. Applying this, two people in Denver must have the same number of hairs on their heads, because no one has more than a million hairs, and Denver has more than a million inhabitants. If we increase the size of the holes, we can claim that if $p + q - 1$ pigeons share two holes, the first hole has p pigeons or the second has q pigeons, at least.

Applications of the various forms of this obvious principle are sometimes not at all obvious, as we now shall see.

Example 3. Every sequence of $n^2 + 1$ distinct integers contains either an increasing subsequence of length $n + 1$ or a decreasing subsequence of length $n + 1$. To see this, let $a_1, a_2, \ldots, a_{n^2+1}$ denote the sequence of integers. Supposing there is no increasing subsequence of length $n + 1$, we will prove there is a decreasing subsequence of that length. For each a_k let i_k be the length of the longest increasing subsequence beginning at a_k. Since all $n^2 + 1$ of the i_k's are between 1 and n, some label, say m, must be used at least $n + 1$ times. The $n + 1$ integers a_k for which $i_k = m$ must form a decreasing subsequence as required.

Example 4. If $a_1, a_2, \ldots, a_n$ are any n integers (not necessarily distinct), then some of the a_i's sum to an integer multiple of n. If we denote the sum of the first k a_i's by S_k, then there are two possibilities. If n divides $S_{m'} - S_m$, for some m' and m, $m' > m$, then $a_{m+1} + a_{m+2} + \cdots + a_{m'}$ is a multiple of n. Otherwise, the n sums S_k are distinct, modulo n, so one must be congruent to 0 and hence divisible by n.

Example 5. A tennis star decides to prepare for an important tournament by playing at least one match a day, but no more than twenty matches altogether, over a period of two weeks. Show that during some set of consecutive days the star must play exactly seven matches. To do this, we denote the number of matches played through the ith day by a_i. Note that $a_1, a_2, \ldots, a_{14}$ is a monotonically increasing sequence of positive integers. Let $b_i = a_i + 7$. Then the b_i's are also clearly monotonic increasing, and the a_i's and b_i's together form a set of 28 positive integers. Since $a_{14} \leqslant 20$, we see $b_{14} \leqslant 27$, and we have 28 integers between 1 and 27. Thus, $a_i = b_j = a_j + 7$, for some i and j, and seven matches are played on days $j + 1$ to i.

EXERCISES

Prove the following statements:

1. In any 5×5 symmetric Latin square A (so $a_{ij} = a_{ji}$, all i, j), each of the numbers 0 to 4 occurs on the main diagonal.

2. If an $8 \times 8 \times 8$ cube is constructed of 128 blocks of size $2 \times 2 \times 1$, then no matter how that is done, it is possible to find a straight line through the cube which passes only along edges of blocks, never striking one head-on. [*Hint:* How many lines are there through the cube, along edges of blocks?]

3. In any set of 11 integers chosen from $1, \ldots, 20$, there must be one that divides another evenly. This is not true for every choice of 10 integers.

4. If n balls are placed in m boxes and $n < \binom{m}{2}$ then at least two boxes contain the same number of balls.

5. If $a_1, a_2, \ldots, a_n$ is a permutation of $1, 2, \ldots, n$ and n is odd, then the product $(a_1 - 1)(a_2 - 2) \cdots (a_n - n)$ is even.

6. If 5 points are chosen in a square of side 2, two of the points are at a distance at most $\sqrt{2}$ from each other.

7. If 5 points are chosen in an equilateral triangle of side 2, two of the points are at a distance at most 1 from each other.

8. A tennis star who plays at least one match each day, but no more than 132 matches altogether, in 77 days must play exactly 21 games in some string of consecutive days.

9.3. RAMSEY'S THEOREM

The pigeonhole principle, even in its many variations, is only a special case of a very deep result known as Ramsey's Theorem. To lead up to that, we introduce some new notation for what we have already covered in the last section.

Let $R = R(n_1, n_2, \ldots, n_k, 1)$ denote the least number so that if any set S with at least R elements is partitioned into k subsets, the ith subset will contain at least n_i elements, for some i (we will explain the number 1 soon). In particular, if the n_i are all 2 then $R(2, 2, 2, \ldots, 2, 1)$ is $k + 1$, because if a set with $k + 1$ elements is divided into k subsets, one of the subsets will contain at least 2 elements. Similarly, $R(p, q, 1) = p + q - 1$. Thus, the R

notation is just a way to write out the various forms of the pigeonhole principle exactly.

In order to generalize the pigeonhole principle, we restate the definition of $R(n_1, n_2, \ldots, n_k, 1)$ in the following rather strange way. Let $R = R(n_1, n_2, \ldots, n_k, 1)$ denote the least number so that if the 1-subsets of any set S with at least R elements are partitioned into k families there is an n_i-subset all of whose 1-subsets are in the ith family of the partition, for some i. This explains the number 1 after the n_i's, and it just says what the previous definition said: the ith subset contains at least n_i elements, for some i. The advantage of the restatement is that it allows an obvious generalization from 1-subsets to r-subsets.

The *Ramsey number* $R = R(n_1, n_2, \ldots, n_k, r)$ is the least number so that if the r-subsets of any set S with at least R elements are partitioned into k families there is an n_i-subset all of whose r-subsets are in the ith family for some i. Unlike the case $r = 1$, in the general case it is not obvious that such an R even exists. We will prove that it does, but first we give some examples from graph theory that illustrate the case $r = 2$.

Consider a complete graph, K_n. Its vertices are elements of an n-set and its edges correspond to all 2-subsets of the n-set. A natural way to indicate a partition of those edges into k sets is to color them with k colors. Thus, $R(n_1, n_2, \ldots, n_k, 2)$ is the size of the smallest complete graph so that if the edges of that graph are colored with k colors, there is an n_i-subset of vertices, all of whose edges are in the ith color class. That is, there is a complete subgraph of size n_i, all of whose edges are i-colored for some i.

The first example of this chapter claimed that in any group of six people, there are three people who are mutual friends or three who do not know each other. If we let the people be the vertices of K_6, and color edges red to indicate friends, blue to indicate strangers, the claim says that every K_6 with red and blue edges has an all-red K_3 or an all-blue K_3. Thus, $R(3, 3, 2) \leqslant 6$. It is easy to color the edges of K_5 with a red pentagon and a blue pentagram (star), so $R(3, 3, 2) > 5$. Thus, we see that $R(3, 3, 2) = 6$. Other small Ramsey numbers can also be calculated.

Example 6. Use graph theory to show $R(4, 3, 2) \leqslant 10$. That is, every 2-colored K_{10} contains a red K_4 or a blue K_3. Since a vertex v of K_{10} is on 9 edges, the pigeonhole principle tells us that either v is on 6 red edges or it is on 4 blue edges. If v is on 6 red edges, then the 6 vertices at the other ends of those edges are vertices of a K_6. If that K_6 contains a red triangle, that triangle plus v makes a red K_4. Otherwise the K_6 contains a blue triangle. In either case, we are done. If v is not on 6 red edges, it must be on 4 blue edges. Those edges go to the vertices of a K_4. If that K_4 has a blue edge, we have a blue triangle. If not, we have a red K_4.

A more complicated analysis can be used to show that $R(4, 3, 2) = 9$. Other known values of $R(m, n, 2)$ are shown in Table 9.1. Note that only very small Ramsey numbers are known; these are not easy problems.

Table 9.1. Ramsey Numbers $R(m, n, 2)$

m \ n	2	3	4	5	6	7
2	2	3	4	5	6	7
3	3	6	9	14	18	23
4	4	9	18			
5	5	14				
6	6	18				
7	7	23				

The major result in Ramsey theory is that Ramsey numbers always exist, even when we cannot calculate them. We state and prove this result first in the case involving a partition of the r-subsets into two families. The general case of a division into k families follows very easily as a corollary.

Theorem 9.1 (Ramsey's Theorem, two families). Given any integers p, q, r with $r \geqslant 1$, $p \geqslant r$, $q \geqslant r$, there exists a number $R(p, q, r)$ such that for any set S with at least $R(p, q, r)$ elements and any division of the r-subsets of S into two families T_1 and T_2, either there is a p-subset A of S, whose r-subsets all lie in T_1, or there is a q-sunset B of S, whose r-subsets all lie in T_2.

To prove the theorem, we prove a series of lemmas. It is understood that for any triple (p, q, r) considered we have $p \geqslant r$, $q \geqslant r$.

Lemma 9.1. $R(p, q, 1) = p + q - 1$, where $p \geqslant 1$, $q \geqslant 1$.

Proof. Since $r = 1$, the r-sets of S are the elements of S themselves. Let the elements of S be divided into two families T_1 and T_2. If $N = |S| \geqslant p + q - 1$, then if $|T_1| \geqslant p$ we can choose A to be a p-set from T_1. If $|T_1| < p$, then $|T_2| \geqslant q$. Hence we can choose B to be a q-set from T_2. On the other hand, if $N = p + q - 2$ then we can divide S into two families T_1 and T_2 such that $|T_1| = p - 1$ and $|T_2| = q - 1$. Choice of the subsets A and B satisfying the conditions of the theorem is not possible. Hence, $R(p, q, 1) = p + q - 1$.

Lemma 9.2. $R(p, r, r) = p$, $R(r, q, r) = q$.

Proof. Let $N = |S| \geqslant p$. If any r-subset belongs to the family T_2 then we can take B to be this subset. If the family T_2 is empty then T_1 is the family of all r-subsets of S. Then any p-subsets of S can be taken as A. If $N < p$, then we can take T_2 to be empty, and the choice is impossible since there is no p-subset of S. Hence $R(p, r, r) = p$. Similarly, $R(r, q, r) = q$.

Thus the theorem is true if $r = 1$ or if one of p and q equals r.

Lemma 9.3. If the theorem is true for every triple $(p^*, q^*, r - 1)$, $p^* \geqslant r - 1$, $q^* \geqslant r - 1$, and for the triples $(p - 1, q, r)$ and $(p, q - 1, r)$, then it is true for (p, q, r). In fact, if we write

$$p_1 = R(p - 1, q, r), \qquad q_1 = R(p, q - 1, r),$$

then

$$R(p, q, r) \leqslant R(p_1, q_1, r - 1) + 1.$$

Proof. Let S be a set with $N = R(p_1, q_1, r - 1) + 1$ elements. Let T_1 and T_2 be two mutually exclusive families of all r-subsets of S. We then have to show that either there exists a subset A of S with p elements such that all r-subsets of A belong to T_1 or there exists a subset B of S with q elements such that all r-subsets of B belong to T_2. Let a_0 be any particular element of S. Let $S' = S - a_0$. Let T'_1 be the family of $(r - 1)$-subsets of S' obtained by deleting a_0 from those r-sets of T_1 that contain a_0, and similarly let T'_2 be the family of $(r - 1)$-subsets of S' obtained by deleting a_0 from those r-sets of T_2 that contain a_0. Then T'_1 and T'_2 are mutually exclusive families of all $(r - 1)$-subsets of S'. This is seen as follows: Let X' be any $(r - 1)$-subset of S'. Then $X = X' + a_0$ is an r-subset of S and therefore belongs to either T_1 or T_2. Hence X' belongs to either T'_1 or T'_2.

Since $|S'| = R(p_1, q_1, r - 1)$ either there exists a subset A' of S' with p_1 elements such that all $(r - 1)$-subsets of A' belong to T'_1 or there exists a subset of B' of S' with q_1 elements such that all $(r - 1)$-subsets of B' belong to T'_2. Suppose the first possiblity is true. Since $|A'| = R(p - 1, q, r)$ either there exists a subset B of A' with q elements such that all r-subsets of A' belong to T_2, in which case B is the required subset, or there exists a subset A_0 of A' with $p - 1$ elements such that all r-subsets of A_0 belong to T_1, in which case we can take $A_0 + a_0 = A$ as our required subset. In the same way we can deal with the second possibility. This proves the lemma.

Lemma 9.4. If the theorem is true for every triple $(p^*, q^*, r - 1)$ and for every triple (p_0, q_0, r), with $p_0 + q_0 \leqslant m - 1$, then it is true for any triple (p, q, r), for which $p + q = m$.

Proof. If $p + q = m$, then by hypothesis the theorem holds for the triples $(p - 1, q, r)$ and $(p, q - 1, r)$. The required result follows from Lemma 9.3.

Lemma 9.5. If the theorem is true for every triple $(p^*, q^*, r - 1)$ then it is true for any triple (p, q, r).

Proof. From Lemma 9.2, $R(r, r, r) = r$. Hence the theorem holds for the triple (r, r, r) with $p + q = 2r$. The required result follows from Lemma 9.4 by induction on $p + q$.

Proof of Theorem 9.1. From Lemma 9.1, the theorem holds for the triple $(p, q, 1)$, that is, for $r = 1$. Theorem 9.1 now follows from Lemma 9.5 by induction on r.

Corollary 9.1 (Ramsey's Theorem, k families). Given any integers $n_1, n_2, \ldots, n_k$ and r with $r \geqslant 1$, $n_i \geqslant r$ (all i), there exists a number $R = R(n_1, n_2, \ldots, n_k, r)$ such that for any set S with at least R elements and any division of the r-subsets of S into k families $T_1, T_2, \ldots, T_k$, there is an n_i-subset of S, whose r-subsets all lie in the family T_i, for some i.

Proof. We prove this by induction, noting that the theorem proves it for $k = 2$. Suppose that it has been shown true for all values of k less than m. To prove it for $k = m$, let S be a set of size at least $R(n_1, R(n_2, n_3, \ldots, n_k, r), r)$ (this expression involves two Ramsey numbers, both of which exist by induction). Then we know from the theorem that there is an n_1-subset of S, all of whose r-subsets lie in the family T_1 (in which case we are done), or there is an $R(n_2, n_3, \ldots, n_k, r)$-subset of S, all of whose r-subsets avoid the family T_1. In the latter case, we are done by induction.

EXERCISES

1. Derive values for $R(p, q, r)$ for all p, q, r up to 3, 3, 3.

2. Use graph theory to show:

(a) $R(4, 4, 2) \leqslant 24$,

(b) $R(3, 3, 3, 2) \leqslant 17$,

(c) $R(3, 3, 3, 3, 2) \leqslant 66$.

3. Give graph constructions that show:

(a) $R(3, 3, 2) \geqslant 6$,

(b) $R(4, 3, 2) \geqslant 8$.

4. Use the proof of Ramsey's Theorem and its corollary to obtain lower bounds for:

(a) $R(5, 5, 2)$,

(b) $R(m, n, 2)$,
(c) $R(4, 4, 3)$,
(d) $R(3, 3, 3, 2)$.

5. Show that $R(m, r, r) = m$.

6. Show that $R(2, 2, 2, \ldots, 2, 1) = k + 1$, if there are k 2's.

9.4. ERDÖS AND SZEKERES' THEOREM

A geometric figure is called *convex* if every line joining two points u and v in the figure lies entirely within the figure. A triangle and an interior point can be considered as determining a quadrilateral (in fact, three different quadrilaterals are possible) that is not convex. Figure 9.2 shows two quadrilaterals, one convex and one not convex.

Three points in the plane necessarily determine a triangle, which is necessarily convex. Four points determine a quadrilateral, which may or may not be convex. Erdös and Szekeres noted that given *five* points in the plane, no three on a line, some four of them form a convex quadrilateral. They used Ramsey's Theorem to prove that given enough points in the plane, convex figures of arbitrarily large numbers of points can be found.

Theorem 9.2. For any given integer $k \geqslant 3$, there exists an integer $n = n(k)$ such that any n points in a plane, no three on a line, contain k points forming a convex k-gon.

Proof. The theorem is trivially true for $k = 3$ with $n(3) = 3$. We need two lemmas to prove the theorem for $k > 3$.

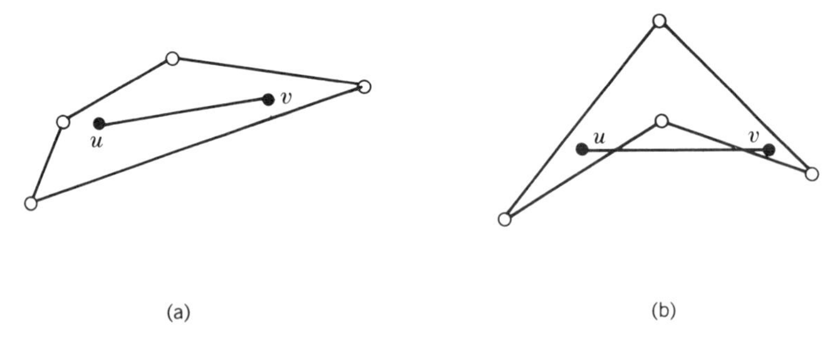

Figure 9.2

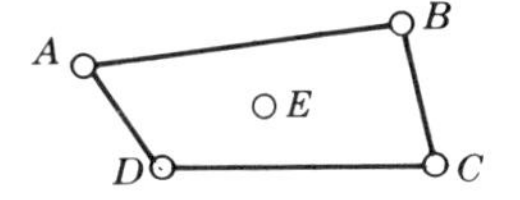

Figure 9.3

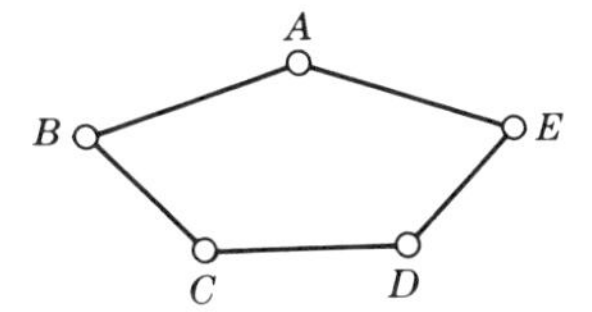

Figure 9.4

Lemma 9.6. $n(4) = 5$.

Proof. We have to show that given any five points in a plane, no three on a line, some four are the vertices of a convex quadrilateral. If the convex hull of the five points is a quadrilateral as in Figure 9.3 or a pentagon as in Figure 9.4 the proof is immediate.

If the convex hull is a triangle ABC and the other two points D and E are inside as in Figure 9.5, then the line DE will intersect two of the sides of the triangle but not the third. Suppose this third side is BC. Then $BCED$ form the vertices of a convex quadrilateral.

Lemma 9.7. If there are k points, no three of which lie on a line, and all the quadrilaterals formed from them are convex, then the points form the vertices of a convex k-gon.

Proof. We shall use induction to prove the lemma. The result is trivial if $k = 4$. Suppose $k \geqslant 5$ and the lemma is true if the number of points is $k - 1$ or less. Let $A_1, A_2, \ldots, A_k$ be k points satisfying the conditions of the theorem. Then $A_1, A_2, \ldots, A_{k-1}$ form a convex $(k - 1)$-gon C_{k-1}. We may suppose that the sides of C_{k-1} are $A_1A_2, A_2A_3, \ldots, A_{k-2}A_{k-1}, A_{k-1}A_1$. If possible, let A_k lie within C_{k-1}, and join A_1 to the other vertices of C_{k-1}. Then since A_k cannot lie on any of the joining lines, A_k must lie inside one of the triangles $A_1A_2A_3, A_1A_3A_4, \ldots, A_1A_iA_{i+1}, \ldots, A_1A_{k-2}A_{k-1}$. Suppose A_k lies within $A_1A_iA_{i+1}$. (Figure 9.6 illustrates the case $k = 7$, $i = 3$). Then

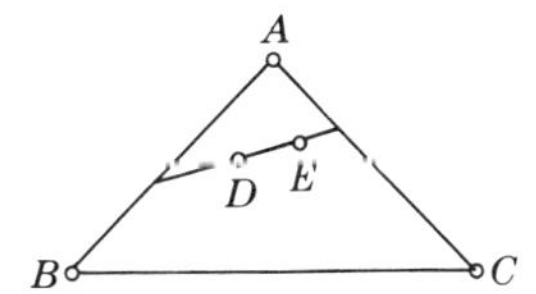

Figure 9.5

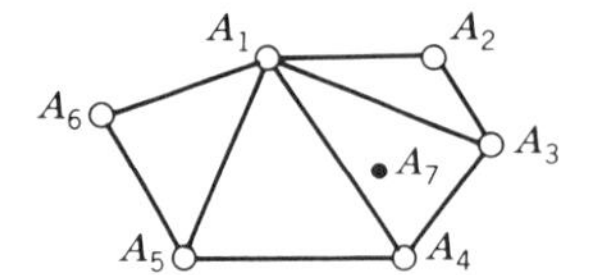

Figure 9.6

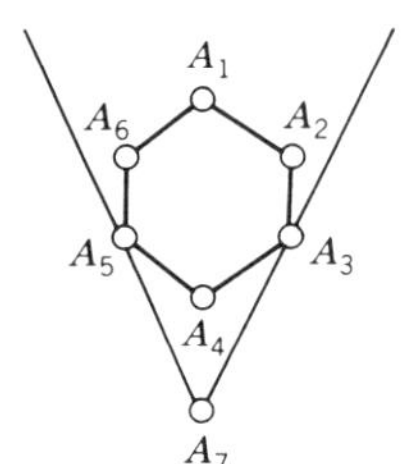

Figure 9.7

$A_1A_iA_{i+1}A_k$ is a concave quadrilateral, which contradicts the hypothesis of the lemma. Hence A_k must lie outside C_{k-1}.

Now joint A_k to the vertices of C_{k-1}. Then there must be two extreme lines A_kA_i and A_kA_j such that C_{k-1} lies within the angle formed by them (Figure 9.7 illustrates the case $k = 7$, $i = 3$, $j = 5$). If A_i and A_j are not consecutive points on the perimeter then there is at least one point A_m between them (for example in Figure 9.7, $m = 4$). In this case $A_kA_iA_mA_j$ form a concave quadrilateral, which is a contradiction. Hence, A_i and A_j are consecutive, say $j = i + 1$ (for example, Figure 9.8 illustrates the case $k = 7, i = 3, j = 4$). Then $A_1A_2 \ldots A_iA_kA_j \ldots A_{k-1}$ is a convex polygon. This proves the lemma.

Now let $n = R(k, 5, 4)$ in the notation of Theorem 9.1. We divide the four subsets of the set S of n points $A_1, A_2, \ldots, A_n$, no three of which lie on a line, into two families T_1 and T_2 such that a 4-subset belongs to T_1 if its four points form a convex quadrilateral and belongs to T_2 if its four points form a concave quadrilateral. Then from Ramsey's Theorem we can either choose a subset of k points from S, such that any four of these points form a convex quadrilateral and hence from Lemma 9.7 a convex k-gon, or there exists a subset of five points of S such that any four of these five points form

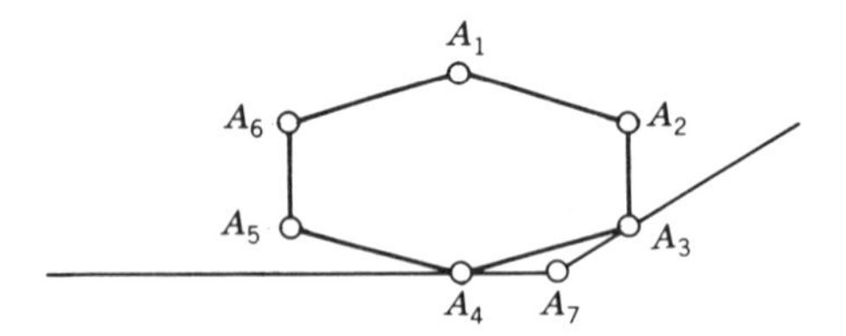

Figure 9.8

a concave quadrilateral. However, from Lemma 9.6 this case cannot arise. Thus there exists a subset of k points of S that form a convex k-gon. This proves Theorem 9.2.

We have shown that $n(k) \leqslant R(k, 5, 4)$. By a different argument Erdös and Szekeres have shown that $n(k) \leqslant \binom{2k-4}{k-2} + 1$.

Now $n(3) = 3 = 2 + 1$, $n(4) = 5 = 2^2 + 1$, and it can be shown that $n(5) = 9 = 2^3 + 1$. From this Erdös and Szekeres conjecture that $n(k) = 2^{k-2} + 1$. However, this conjecture is still unsettled.

The following variation of Erdös and Szekeres Theorem is also unsettled. For any integer $k \geqslant 3$, does there exist an integer $m = m(k)$ such that any m points in a plane, no three on a line, contain k points forming a convex k-gon with none of the other $m - k$ points in its interior? It is easy to see that $m(3) = 3$ and $m(4) = 5$, but for $k > 4$ even the existence of $m(k)$ is in doubt.

EXERCISES

1. Show that $n(5) \geqslant 9$ by displaying a set of eight points in the plane, no three on a line, that determine no convex pentagon.
2. Show that if seven points are chosen on a regular hexagon of side 1, two of the points must be within distance 1 of each other. Is this also true if six points are chosen?
3. Show that if nine lattice points (with integers as coordinates) are chosen in 3-space, then the midpoint of the line joining some two of the points is also a lattice point.

9.5. PARITY

We used the odd–even aspect of our number system heavily during our work on eulerian graphs in Chapter 4. Many other existence results in combinatorics depend on the pigeonhole principle in a way involving even and odd numbers. We present several examples here.

Example 7. A domino covers exactly two squares of a standard, 8×8 chess board. Is it possible to cover the whole chess board with dominoes? What if a single corner square is removed from the board? What if two opposite corner squares are removed from the board?

The first two answers are obviously yes and no, respectively. It is easy to think of many ways to pave an 8×8 board with dominoes. Also, if one square is removed, only 63 squares remain, and dominoes must cover an even number of squares. The third question is not quite as easy. Removing opposite corners leaves 62 squares, so it should be possible to cover the rest of the board with 31 dominoes. If you try it, however, you will find it cannot be done. To prove it cannot be done, we use a parity argument. Note that the squares of a chess board are colored alternately, say red and black, and opposite corners are the same color. Removing a pair of opposite corners leaves 30 red and 32 black squares, say. Since a domino clearly must cover exactly one red and one black square, no covering is possible.

Example 8. Beginning with 5 coins lying heads up on a table, is it possible to find a sequence of moves that end with all coins heads down if a move consists of turning over any 4 coins? What if we begin with 6 coins?

It turns out that 5 coins cannot be turned over by any sequence of moves. To show that, we note that each move preserves the parity, odd or even, of the number of coins that are heads up. Thus, since we start with 5 coins heads up, we cannot finish with 0 coins heads up. If we begin with 6 coins, however, it is not difficult to find a sequence of four moves that ends with all coins tails up.

Example 9. For which values of m and n is the complete bipartite graph $K_{m,n}$ hamiltonian?

We claim that $K_{m,n}$ is hamiltonian only when $m = n \geqslant 2$. Note that any hamiltonian cycle must visit the parts of $K_{m,n}$ containing m points and n points alternately. Thus, if $m \neq n$, there cannot be a hamiltonian cycle. If $m = n$, a cycle is easy to find explicitly.

Example 10. Suppose we are given a 6×6 chess board, paved with dominoes. Then we claim it is always possible to find a line all the way from one side of the board to the other which crosses no domino. Figure 9.9 shows one possible paving and such a line

To prove that there is always an unblocked line, note first that the number of dominoes crossing any line is even, because the number of squares on each side of the line is even. Note also that there are 10 lines to be blocked and 18 dominoes. Since every blocked line is blocked by at least two dominoes, and no domino blocks more than one line, the pigeonhole principle applies: We have too few dominoes to block the lines!

The next example requires some knowledge of odd and even permutations. Every permutation of the integers 1 to n can be obtained by a series of switches. For example, the following list shows a series of four switches

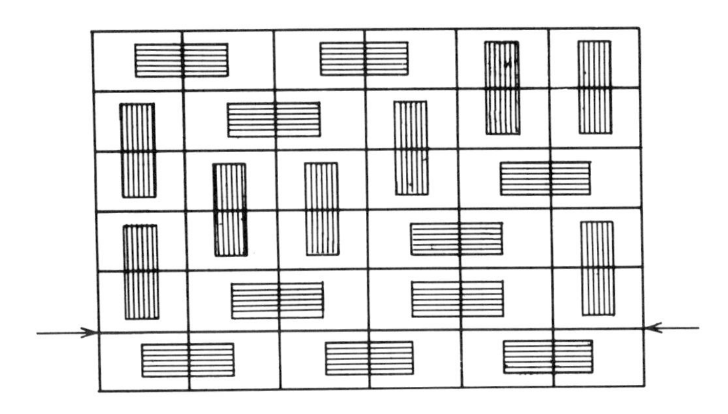

Figure 9.9

leading from the natural ordering 1 2 3 4 5 6 7 8 to the ordering 5 3 6 8 1 2 7 4. The italics indicate which numbers are to be switched next.

1 2 *3* 4 5 *6* 7 8
1 2 6 *4* 5 3 7 *8*
1 *2* 6 8 5 *3* 7 4
1 3 6 8 *5* 2 7 4
5 3 6 8 1 2 7 4

Obviously there are other switching sequences that would take us from 1 2 3 4 5 6 7 8 to 5 3 6 8 1 2 7 4, but we claim that every such sequence would contain an even number of switches, just as the one we show here has four. A permutation obtained from the natural one by an even (odd) number of switches is called an *even (odd) permutation.* To show that our claim on the parity of the number of switches is true, we introduce another concept. For any permutation p of 1 to n the number of *inversions* $i(p)$ is the number of pairs x, y with x preceding y in p and $x > y$. The permutations listed above in our switching list have 0, 5, 10, 11, and 14 inversions, respectively.

Note that with each switch, the parity of $i(p)$ changes. That is a general phenomenon, as we now prove. Say x and y are to be switched, and x comes before y in the permutation. If x and y are adjacent in the permutation, clearly switching x and y increases or decreases the number of inversions as x is less than or greater than y. Thus, if x and y are adjacent on p, switching x and y changes the parity of $i(p)$. If x and y are not adjacent, suppose that $x < y$ (the case $x > y$ is similar and left for the exercises). We can concentrate our attention on inversions involving x, y and elements in between, because other inversions do not change when x, y are switched. Suppose that among the numbers between x and y in the permutation p, a are less than x, b are between x and y, and c are greater than y. Then before x and y are switched, x is in an inversion with a elements between x and y, and y is in c such inversions. After x and y are switched, x is in $a + b$ such inversions, and y is in $b + c$, and we also have one new inversion in the pair yx. Thus the change in the number of inversions is $(a + b + b + c + 1) - (a + c) = 2b + 1$, which is odd. Thus, a switch changes the parity of $i(p)$.

It then follows that our definition of even and odd permutations is valid, because the parity of $i(p)$, which is determined by p, specifies whether we have made an even or an odd number of switches to reach p from the natural ordering. This theory has a nice application to a famous puzzle.

Example 11. The Fifteen Puzzle, shown in Figure 9.10(a), consists of 15 numbered squares, free to slide in a 4×4 frame. The object of the puzzle is to move the squares so that another configuration is obtained. In fact, some patterns, including Figure 9.10(b), are impossible.

To see that, think of the first pattern as the natural ordering of 1 to 16 (call the blank 16). Then a move is just a switch of 16 and another number. But the permutation for the second pattern is $p = 16\ 1\ 2\ 3\ 4\ 5\ 6\ 7\ 8\ 9\ 10\ 11\ 12\ 13\ 14\ 15$ and $i(p) = 15$. Thus p is the result of an odd number of switches of the blank square. But the blank square starts and ends on a red square of the 4×4 chess board in the change from (a) to (b) so it must make an even number of moves. Thus p is impossible.

In general a permutation p will determine the parity of the number of moves in two ways: obviously from the red–black chess board positions of the blank square before and after and less obviously from the parity of $i(p)$. If these parities are different, p is an impossible position. It is not very difficult to show that if those parities are the same, p is obtainable by some sequence of moves.

1	2	3	4
5	6	7	8
9	10	11	12
13	14	15	

(a)

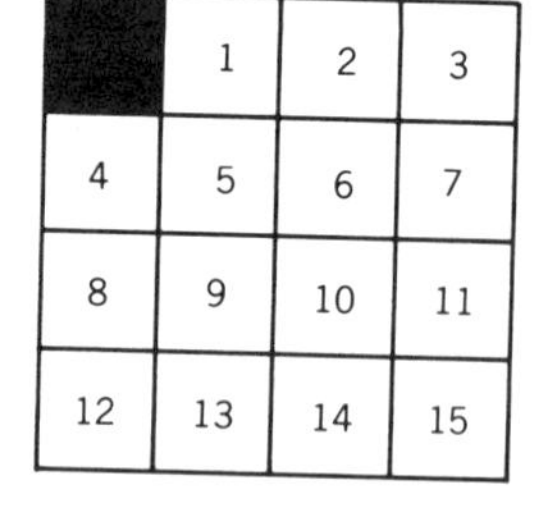

(b)

Figure 9.10

EXERCISES

1. Use the following diagram to show that a chess board missing any two squares of the same color can be covered with dominoes.

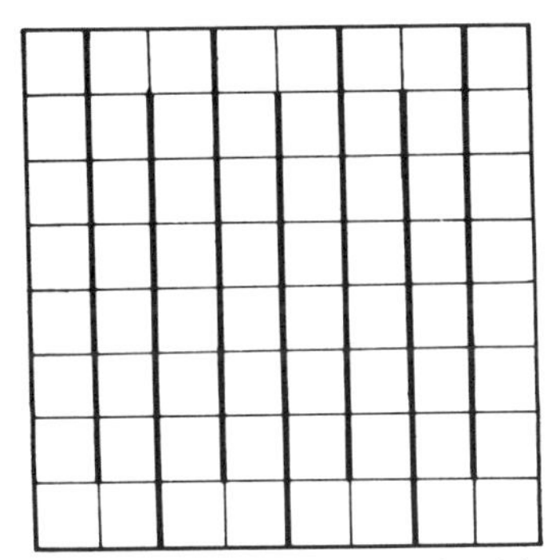

2. Show that if a 5×6 chess board is paved with dominoes, a line can be found that passes from some side of the board to the opposite side, crossing no domino. Is the same true of a 4×7 board?

3. A $20 \times 20 \times 20$ cube is constructed of 2000 $2 \times 2 \times 1$ blocks. Show that it is always possible to find a straight line through the cube which passes only along edges of blocks, never striking one head-on.

4. Find $i(p)$ if p is the permutation:
   (a) 3 1 4 5 2 6 7 8,
   (b) 8 5 2 1 3 4 7 6.

5. Show that if $x > y$ and x precedes y in the permutation p, then switching x and y changes the parity of $i(p)$.

6. Which of the following permutations of 1 2 3 4 5 6 7 8 are even?
   (a) 3 1 4 5 2 6 7 8,
   (b) 8 5 2 1 3 4 7 6,
   (c) 8 7 6 5 4 3 2 1.

7. Which of the following configurations of the Fifteen Puzzle can be obtained from Figure 9.10(a)?

1	5	9	13
2	6	10	14
3	7	11	15
4	8	12	

(a)

1	3	5	7
9	11	13	15
2	4	6	8
10	12	14	

(b)

1	3	6	10
2	5	9	13
4	8	12	15
7	11	14	

(c)

8. A $3 \times 3 \times 3$ cube contains 27 unit cubes, so such a cube with a hole in the center contains 26 unit cubes. Can such a cube-with-hole be assembled from 13 blocks each $2 \times 1 \times 1$?

9. Show that four copies of the "T" shown below can be assembled to form a 4×4 square. Can five copies be assembled into a 4×5 rectangle? [*Hint:* Color the board in chess board fashion.]

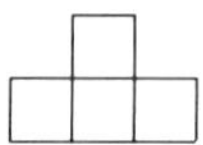

10. Give a simple method for deciding whether the points w, x, and y are enclosed by the closed curves in (a), (b), and (c) below.

9.6. REMARKS

The pigeonhole principle has its origin in the folklore of mathematics, although it is sometimes also called the Dirichlet Drawer Principle, after a famous French mathematician.

Ramsey's Theorem appeared in 1930 and Erdös' and Szekeres' application to convex polygons appeared in 1935, but most of the development of Ramsey theory has occurred in the last twenty years. Today, both calculation of various Ramsey numbers and generalizations of the theory are receiving a good deal of attention.

Parity arguments are a favorite tool in many problems of recreational mathematics, as the examples we give suggest, but they are also important in serious mathematics.

For an extensive surve of Ramsey theory, see the book by Graham, Rothschild, and Spencer [1]. For more about even and odd permutations, switchings, and applications to sorting algorithms, see the monumental work of Knuth [2]. A thorough but elementary treatment of the Fifteen Puzzle can be found in Stein [3].

[1] R. L. Graham, B. L. Rothschild, J. H. Spencer, *Ramsey Theory*, Wiley, New York, 1980.

[2] Donald E. Knuth, *The Art of Computer Programming*, Vol. 3, Addison-Wesley, Reading, 1973.

[3] S. K. Stein, *Mathematics, the Man-Made Universe*, Freeman, San Francisco, 1976.

CHAPTER 10

Optimization

10.1. INTRODUCTION

Combinatorics has traditionally focused on questions of existence, enumeration, and construction. We might begin by asking whether a finite structure with given properties exists. If the answer is an emphatic or obvious yes, the next natural problem would be to enumerate such structures. If, on the other hand, existence is a difficult question, one might be more interested in methods for constructing examples of such structures.

Recently, a fourth line of inquiry has been added to these three. Combinatorial optimization attempts to find the "best" structure with specified properties. In optimization problems it is usually obvious that a solution exists, and there are far too many solutions to encourage enumeration. Construction of a solution is important, but it must be a "good" solution, by some measure. The recent blossoming of the optimization branch of combinatorics is largely due to the development of computers. Although the methods used are made as efficient as possible, it may often be necessary to compare thousands of choices before finding the best one, a task that only computers can do.

Example 1. Seven boys are acquainted with ten girls as follows: Al knows Barbara, Debby, and Helen; Bob knows Flo and Debby; Clark knows Barbara and Debby; Dan knows Flo, Jan, and Inez; Evan knows Barbara and Helen; Fred knows Flo and Helen; and Glen knows Alice, Carol, Emily, and Gloria. What is the maximum number of boys who can have dates at the same time? In particular, is it possible to pair the seven boys with seven girls so that each boy is paired with a girl he knows?

It is not difficult to find a pairing that gives six boys partners, but a pairing for all seven is not possible. Perhaps the easiest way to see that is to note that Al, Bob, Clark, Evan, and Fred only know a total of four different girls.

Example 2. Is it possible to extend the following rectangle to a 7×7 Latin square?

$$\begin{matrix} 0 & 1 & 2 & 3 & 4 & 5 & 6 \\ 4 & 2 & 1 & 0 & 6 & 3 & 5 \\ 3 & 0 & 4 & 5 & 1 & 6 & 2 \\ 6 & 4 & 5 & 2 & 3 & 0 & 1 \end{matrix}$$

Perhaps a better way to phrase the question, in view of our optimization theme, is: What is the maximum number of additional rows that can be added, preserving the Latin character of the rectangle? In fact, there are three rows that can be added, to make a Latin square, namely:

$$\begin{matrix} 1 & 3 & 6 & 4 & 5 & 2 & 0 \\ 2 & 5 & 3 & 6 & 0 & 1 & 4 \\ 5 & 6 & 0 & 1 & 2 & 4 & 3 \end{matrix}$$

Example 3. Consider the graph of Figure 10.1. Clearly there is a path from x to y that goes along the top of the graph, and another x–y path that goes along the bottom, and those two paths are disjoint except for the vertices x and y. It is not difficult to find a third path from x to y that avoids the top path and the bottom one. Thus, there is a set of three x–y paths that are disjoint except for x and y. Is this the largest such set of paths, or can four disjoint x–y paths be found?

In fact, there cannot be four disjoint paths from x to y. To see that, note that the vertices a, b, and c in Figure 10.1 effectively separate x from y. Thus, every path from x to y must contain one of a, b, or c. In any set of four or more x–y paths, the pigeonhole principle tells us that two of the paths in the set must pass through a or b or c, so the paths are not disjoint.

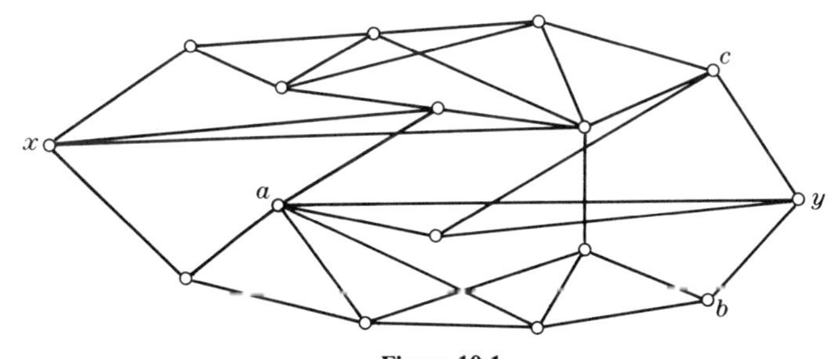

Figure 10.1

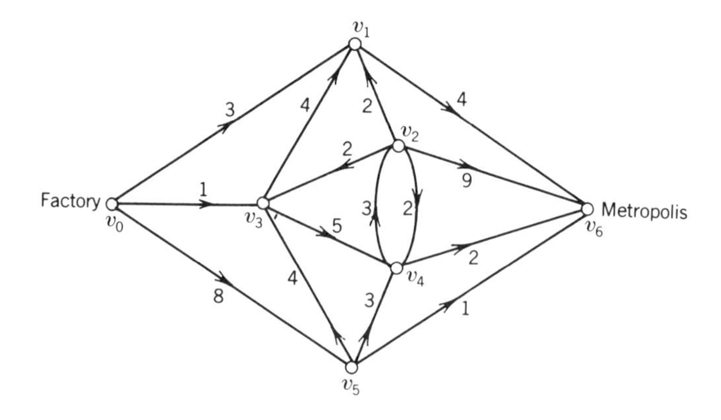

Figure 10.2

Example 4. Bigco manufactures widgets in a single factory and ships them all to market in Metropolis through various shipping channels (trucks, rail, barges, etc.). Figure 10.2 shows these various channels and their capacities in thousands of widgets per day. The arrows on the various channels indicate that widgets can only be transferred in the direction shown. What is the largest possible number of widgets that can be transferred to Metropolis each day, and how should widgets be moved in each channel to achieve that number?

Experimentation shows that 10,000 widgets can be shipped each day and in fact that is the maximum capacity, although that is not easy to prove without some theory.

In this chapter we develop the optimization tools needed to solve problems like those presented in these examples.

EXERCISES

1. List a way to pair six boys and six girls in Example 1.

2. Is it possible to match all seven boys with girls in Example 1 if we first introduce Al to Flo, Evan to Debby and Flo, and Dan to Barbara and Debby?

3. Replace the first row of the Latin rectangle of Example 2 by 2 6 3 4 5 1 0 to form a new 4×7 Latin rectangle L. Then complete L to a Latin square.

4. Complete the following partial Latin square:

$$\begin{array}{cccccc} 0 & 1 & 2 & 3 & 4 & 5 \\ 1 & 5 & 3 & 4 & 2 & 0 \\ 3 & 2 & 5 & 0 & 1 & 4 \\ 2 & 4 & & & & \\ 4 & 3 & & & & \\ 5 & 0 & & & & \end{array}$$

5. (a) Find the maximum number of vertex-disjoint paths from s to t in each of the following graphs and show that it is the maximum.

(b) Do the same for edge-disjoint paths, which may share vertices but not edges.

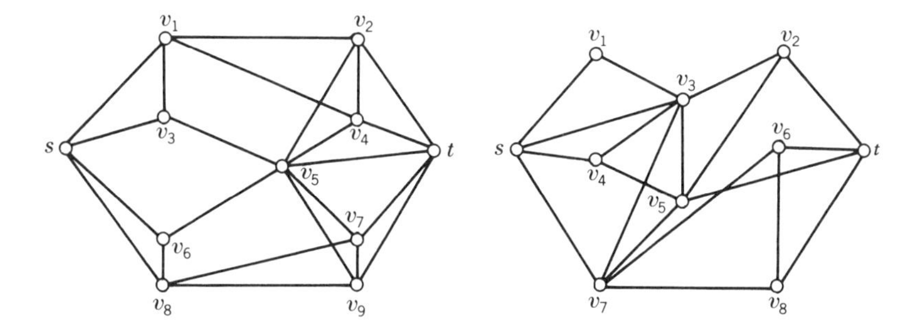

6. (a) Describe a way to transport 10,000 widgets daily, in Figure 10.2, by specifying the amount of widgets to send along each channel $v_i v_j$.

(b) Find a set of channels with total capacity 10,000 that, if blocked, would completely cut off the flow of widgets from the factory to Metropolis.

10.2. THE MARRIAGE THEOREM

Example 1 in the last section belongs to an important general class of problems variously known as matching problems, assignment problems, or systems of distinct representatives. The major theorem in the area can be stated in terms of marriages, and we adopt that intuitively appealing terminology here.

Given a set M of men and a set W of women, together with a set S of acceptable pairs (m_i, w_j), $m_i \in M$, $w_j \in W$, the *marriage problem* is to determine whether or not it is possible to pair off every man m_i with a woman w_j so that every man is happy [i.e., $(m_i, w_j) \in S$] and no one is married twice. Note that the problem is not symmetrical; we want to marry off all the men with women they like, but left-over women do not count. Obviously a problem written by a male mathematician! Note also that the problem has an obvious interpretation in terms of bipartite graphs, with the pairs (m_i, w_j) forming edges of the graph.

Thinking back to Example 1, we realize that a necessary condition for having a solution to the marriage problem is that every set of k men must know, collectively, at least k women. For example, if some man knows no woman the problem cannot be solved. If two men each know only one woman, and she is the same woman, one of the men cannot be married. In Example 1, there was a set of five men who knew, together, only four women. Thus, no marriage would be possible for all five men in that case. The surprising thing is that this obvious necessary condition is also sufficient to assure that the marriage problem can be solved. This marriage theorem was proved by Phillip Hall in 1935.

Theorem 10.1 (Marriage Theorem). A necessary and sufficient condition for the marriage problem to have a solution for n men is that every set of k men must, together, know at least k women, for each k, $1 \leqslant k \leqslant n$.

Proof. The condition is clearly necessary, as we have already explained. To prove it sufficient we will use induction on n, the number of men.

For $n = 1$, we have one man and the condition assures that he can marry, so the marriage problem has a solution. Suppose then that we have shown that for any set of $n - 1$ or fewer men, the condition of the theorem assures that those men can be married. We need to show that the same is true for a set of n men.

We break the proof into two cases. Suppose first that every set of k of the men, $1 \leqslant k < n$, knows together at least $k + 1$ women. That is, every set of k men satisfy the condition of the theorem with a woman left over. In this case we pair any one man with any one woman he knows, reducing us to a marriage problem with only $n - 1$ men Because there was always an extra woman before, we know that any k men must still know at least k women, and we can marry them off by our induction assumption.

So suppose, as the second case, that there is some set of k men, $k < n$, who know, together, exactly k women. Obviously, we must marry those men to those women, so we do that, leaving $n - k$ men. But any collection of j of those $n - k$ men, $1 \leqslant j \leqslant n - k$, must know at least j of the remaining women. That is so because otherwise these j men, together with the k men we married off, would together know less than $j + k$ women, which is impossible. Thus the $n - k$ men who are left satisfy the hypothesis of the theorem and can be married off by our induction assumption.

Although the theorem is stated in rather colorful and playful terms, it has many useful consequences, which we now investigate. A *matching* in a graph G is a subgraph that is regular of degree 1. That is, it consists entirely of isolated edges. If G is bipartite, with parts V_1 and V_2, then clearly the edges of a matching all have one vertex in V_1 and one in V_2. If all vertices of V_1 are contained in the matching, it is called a *complete matching from* V_1 *to* V_2. For any set A of vertices, we denote the set of vertices adjacent to at least one vertex of A by $N(A)$. With all of this terminology, we can state the graph-theoretic form of the marriage theorem.

Theorem 10.2. If G is a bipartite graph with parts V_1 and V_2 then there is a complete matching from V_1 to V_2 if and only if $|N(A)| \geqslant |A|$ for all subsets A of V_1.

Proof. Interpreting V_1 as men, V_2 as women, and the edges of G as the set of acceptable pairs, we see that a complete matching is exactly a solution for the marriage problem, so this is just a restatement of the marriage theorem.

Example 5. The graph shown in Figure 10.3(a) satisfies the conditions of Theorem 10.2. The easiest way to show that is to find the complete matching from V_1 to V_2, shown with darkened edges! Probably the greatest utility of Theorem 10.2 lies in proving that a given graph does *not* have a complete matching. For example, in Figure 10.3(b), the vertices v_1, v_4, v_5, v_6 form a set of four vertices that, together, are adjacent to only three

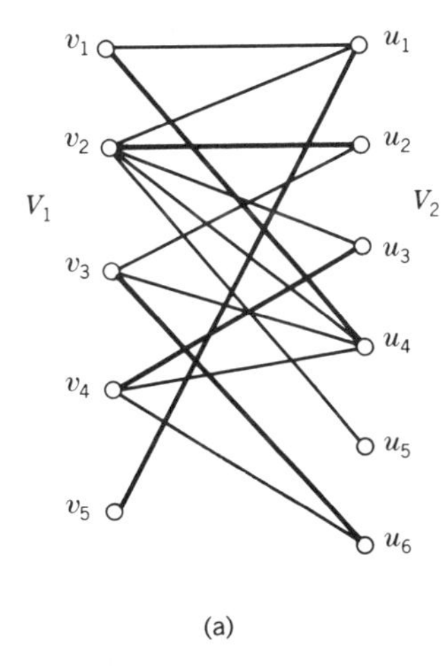

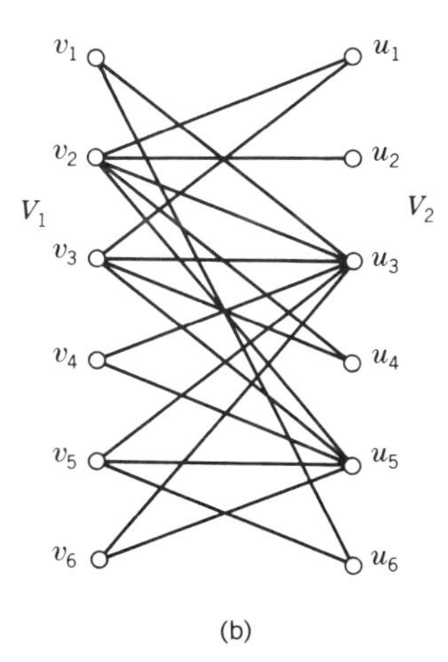

Figure 10.3

vertices in V_2. Thus, the theorem assures us that no complete matching from V_1 to V_2 is possible.

The marriage theorem can be stated in terms of sets. Let $\{S_1, S_2, \ldots, S_n\}$ be a family of subsets of a finite set S. We say that $x_1, x_2, \ldots, x_n$ is a *system of distinct representatives* (SDR) if $x_i \in S_i$ $(i = 1, 2, \ldots, n)$ and $x_i \neq x_j$ if $i \neq j$. We do not require that the subsets S_i are disjoint, or even distinct from each other. Phillip Hall originally proved the marriage theorem in the following form.

Theorem 10.3. (Hall's Theorem). A family $\{S_1, S_2, \ldots, S_n\}$ of subsets of a finite set S has a system of distinct representatives if and only if for $k = 1, 2, \ldots, n$, the union of any k of the sets S_i contains at least k elements of S.

Proof. Again we interpret in terms of marriages, letting the women be the elements of S and the set S_i denote those women who are acceptable to man i. Then a system of distinct representatives is a way to choose for each man, from 1 to n, a wife from his set of acceptable women. The condition on unions of the sets S_i is clearly the same as the condition given in the marriage theorem, so it is equivalent to the existence of an SDR.

Example 6. Let $S = \{1, 2, 3, 4, 5, 6\}$ and say the subsets are $S_1 = \{1, 4\}$, $S_2 = \{1, 2, 3, 4, 5\}$, $S_3 = \{2, 4, 6\}$, $S_4 = \{3, 4, 6\}$, $S_5 = \{1\}$. Then a system of distinct representatives is $x_1 = 4$, $x_2 = 2$, $x_3 = 6$, $x_4 = 3$, $x_5 = 1$. If the subsets are $T_1 = \{3, 6\}$, $T_2 = \{1, 2, 3, 4, 5\}$, $T_3 = \{1, 3, 4, 5\}$, $T_4 = \{3, 5\}$, $T_5 = \{3, 5, 6\}$, $T_6 = \{3, 5, 6\}$, then there is no SDR, because $T_1 \cup T_4 \cup T_5 \cup T_6$ contains only three elements, namely 3, 5, and 6. In fact, these two examples are just the graphs of Figure 10.3(a) and (b), expressed in terms of sets.

The conditions of Theorems 10.1, 10.2, and 10.3 are rather difficult to check. A simple theorem, which we state first in terms of marriages, says that if each man knows enough women, and no woman in known by too many men, all of the men can marry.

Theorem 10.4 Given n men and m women, if every man knows at least k women and every woman is known by at most k men, then all of the men can marry a woman they know.

Proof. Say S is any set of men. Then since each man knows at least k women, the number of man–woman pairs involving men in S is at least $k|S|$. How many different women are involved in those man–woman pairs?

Each woman can be involved in at most k of those pairs, so by the pigeonhole principle there must be at least $k|S| \div k = |S|$ women involved. Thus the men in S know at least $|S|$ women. Since S was an arbitrary set of men, the condition of the marriage theorem is met and all of the men can marry.

Example 7. Ten professors are asked, in order of seniority, to choose their three favorite classes from a long list. As they express preferences, each class is removed from the list as soon as three professors have chosen it. Show that each professor can be assigned a favorite class to teach.

After all the choices are expressed, each professor is matched with three classes and each class with at most three professors. Applying Theorem 10.4, we can "marry" each professor to a favorite class.

We restate Theorem 10.4 in terms of SDR's for future reference.

Corollary 10.1 Suppose an n-set S has subsets $S_1, S_2, \ldots, S_n$ and that each element of S is in at most k of the subsets and each subset contains at least k elements for some $k > 0$. Then the family $S_1, S_2, \ldots, S_n$ has an SDR.

EXERCISES

1. Show that the following men cannot all get married: Bob knows Carol and Flo; Dave knows Alice, Barb, Carol, Debby, and Emily; Frank knows Alice, Carol, Debby, and Emily; Henry knows Carol and Emily; Ned knows Carol, Emily, and Flo; Stu knows Carol, Emily, and Flo.

2. Men A, B, C, D, and E are qualified for jobs 1, 2, 3, 4, 5, 6, and 7 as shown by the following diagrams. Show that in one case they can be assigned so they all have jobs, but in the other case they cannot.

	1	2	3	4	5	6	7
A	—	—	√	√	√	√	√
B	√	—	—	—	—	√	—
C	√	—	—	—	—	√	—
D	√	—	√	—	—	—	—
E	√	—	√	—	—	√	—

	1	2	3	4	5	6	7
A	√	√	—	√	—	—	—
B	—	—	√	√	—	√	—
C	√	—	√	√	—	—	—
D	√	—	—	√	—	—	—
E	—	√	—	—	√	—	√

3. The bold edges in the following graphs G_i form edge sets S_i.
   (a) Which sets S_i are matchings?
   (b) Which graphs G_i are bipartite?

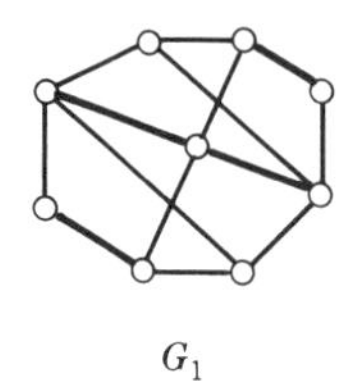

G_1

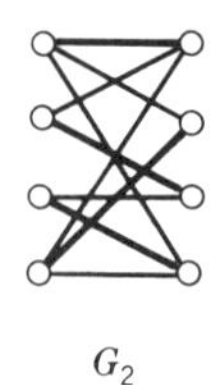

G_2

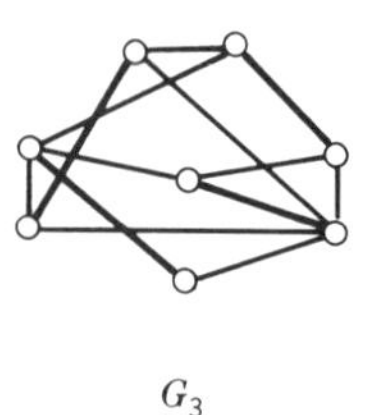

G_3

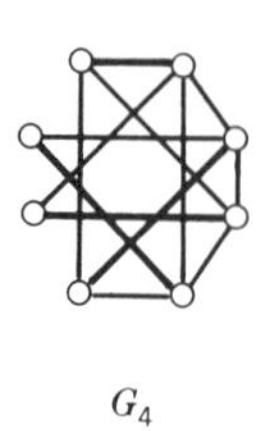

G_4

(c) Which sets S_i in bipartite graphs G_i are complete matchings from some part to the other part?

4. Show that if the graph of Figure 10.3(a) is modified by deleting the edge v_1u_4, the resulting graph has no complete matching from V_1 to V_2.

5. Show that if the graph of Figure 10.3(b) is modified by adding an edge v_1u_1, the resulting graph has a complete matching from V_1 to V_2.

6. Let $S = \{1, 2, 3, 4, 5, 6\}$. For each of the following families of subsets S_i, either find an SDR or show that none can exist.

(a) $S_1 = \{1, 2, 3\}, S_2 = \{2, 3, 5\}, S_3 = \{3,5\}, S_4 = \{1, 2, 3\}, S_5 = \{1,3,5\}$.

(b) $S_1 = \{1, 3, 5\}, S_2 = \{2, 4, 6\}, S_3 = \{1, 2\}, S_4 = \{2, 3, 4\}, S_5 = \{1, 2, 5\}$, $S_6 = \{1, 5\}$.

7. On the tropical island of Melo, social custom dictates that young women are sheltered from birth. Each girl is introduced, in her sixteenth year, to the three boys her parents have selected as best for her. There is general agreement among parents about the desirable mates, so boys who are selected at all are invariably selected by at least three families for introduction to their daughters. Show that in Melo society it is always possible for all the selected boys to marry a girl to whom they have been introduced.

8. The Off-Street Employment Agency has 11 job openings and many candidates for employment, each qualified for three of the jobs. The Agency selects a group of candidates that includes three people qualified for each job. How many candidates have been selected? Can the jobs be filled from those candidates?

9. Prove that if G is a bipartite graph, regular of degree $d > 0$, then G has a complete matching.

10. Show in two ways that there is an SDR for the following family of sets. $S_1 = \{1, 2, \ldots, r), S_2 = \{2, 3, \ldots, r + 1), \ldots, S_i = \{i, i + 1, \ldots, r + i\}$, $\ldots, S_n = \{n, n + 1, \ldots, r + n\}$. First, find an SDR. Second, quote a theorem that guarantees an SDR.

11. For each of the following collections based on the S_i of exercise 10, state when, in terms of n and r, there is an SDR:

(a) $T_i = S_i$, $k = 1, \ldots, n$; $T_{n+1} = S_1$ (family of size $n + 1$).
(b) $R_i = S_i$, $i = 1, \ldots, n$; $R_{n+i} = S_i$, $i = 1, \ldots, n$ (family of size $2n$).

10.3. APPLICATIONS

There are several mathematical consequences of the theorems on marriages, matchings, and SDR's in the last section. Our first result says that it is impossible to go wrong when building a Latin square.

Theorem 10.5. Any $r \times n$ Latin rectangle with $r < n$ can be extended to an $n \times n$ Latin square.

Proof. Let S be the set containing $0, 1, 2, \ldots, n - 1$, and let S_i be the set of elements in S that do not appear in column i of the Latin rectangle for $i = 0$ to $n - 1$. Then clearly each S_i contains $n - r$ elements of S. Furthermore, each element of S has appeared exactly r times so far in the Latin rectangle and in exactly r columns and thus is contained in exactly $n - r$ of the sets S_i. Applying Corollary 10.1, we obtain an SDR $x_0, x_1, \ldots, x_{n-1}$ for the sets S_i. But by the definition of those sets, this is a satisfactory next row for the Latin rectangle, extending it to $r + 1$ rows. Repeating the process, we can eventually obtain a Latin square.

Next we explore the consequences of Theorem 10.4 for graphs. A *1-factor* of a graph G is a spanning subgraph that is regular of degree 1. Thus it is merely a collection of edges that are disjoint and cover all of the vertices (what we have previously called a complete matching). A *1-factorization* is a decomposition of all edges of the graph into a collection of 1-factors, using no edge twice. The idea of 1-factors arises naturally in setting up schedules for sporting events, because each 1-factor specifies a way to pair up opponents for a single round of competition.

Example 8. Six teams are in the Junior Flyball league. Obviously, each team needs to play the other five exactly once. Show that it is possible to set up a schedule that completes all necessary games on five Saturdays.

We can think of the six teams as vertices of the complete graph K_6

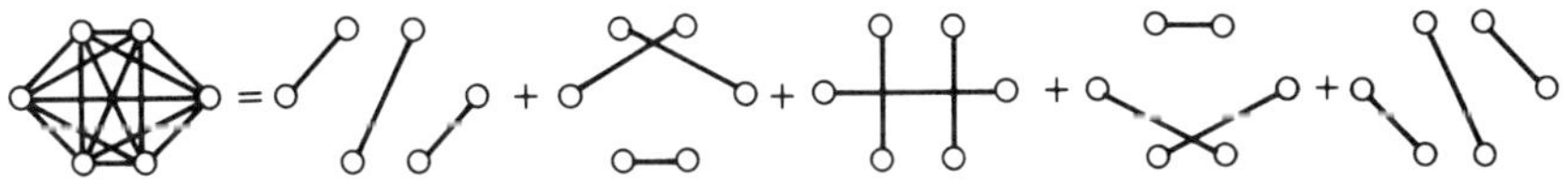

Figure 10.4

and indicate each week's matches by a 1-factor. Then the claim is that K_6 is 1-factorable. Figure 10.4 proves that claim.

In fact, K_n is 1-factorable for every even n (see exercise 3).

Theorem 10.6. Regular bipartite graphs are 1-factorable.

Proof. Suppose G is regular of degree k. Identifying V_1 as men and V_2 as women, with edges indicating marriageable pairs, the conditions of Theorem 10.4 are clearly met and it is possible to match up the vertices in V_1 with the vertices of V_2, using a 1-factor. If those edges are deleted from G, clearly a bigraph that is regular of degree $k - 1$ is obtained and the process can be repeated. Altogether k distinct 1-factors can be found in this way, so we have a 1-factorization of G.

Example 9. Let G be the graph in Figure 10.5. It satisfies the conditions of the theorem.

Three 1-factors into which G can be decomposed are shown in Figure 10.6. First G_1 was found, then G_2 was found in $G - G_1$, and of course G_3 is just $G - G_1 - G_2$.

A final area of application is matrix theory. A matrix is said to be a *zero–one matrix* if each of its elements is zero or one. A *permutation matrix* is defined to be a square zero–one matrix such that each row and each column has a single unity and all the other elements are zero. For example,

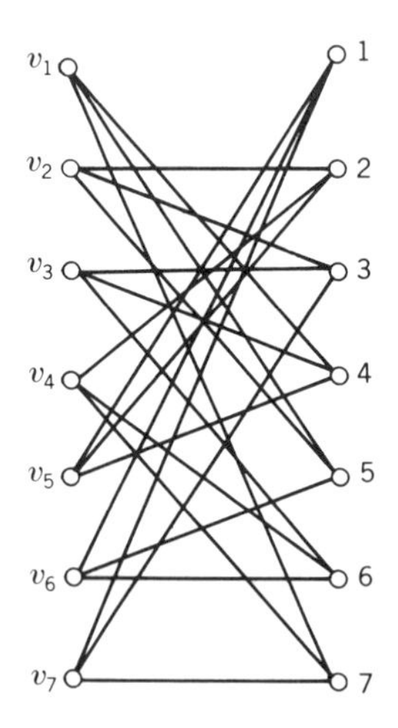

Figure 10.5

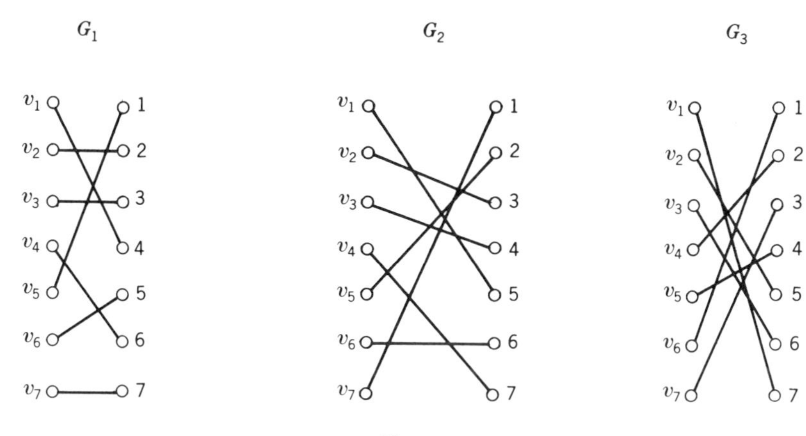

Figure 10.6

$$P = \begin{bmatrix} 1 & 0 & 0 & 0 \\ 0 & 0 & 1 & 0 \\ 0 & 0 & 0 & 1 \\ 0 & 1 & 0 & 0 \end{bmatrix}$$

is a permutation matrix of order 4. It corresponds to the permutation 1 3 4 2 of the number 1 2 3 4. In general a permutation $a_1 a_2 \ldots a_n$ of 1, 2, ..., n will be represented by a permutation matrix in which the ith row has a unity in the a_i^{th} column and has zeros elsewhere ($i = 1, 2, \ldots, n$).

Theorem 10.7. Any $n \times n$ zero–one matrix A such that the sum of each row and each column is k can be written as the sum of k permutation matrices.

Proof. Let $A = (a_{ij})$ be the given matrix. Then $a_{ij} = 0$ or 1, and

$$\sum_{i=1}^{n} a_{ij} = k, \qquad \sum_{j=1}^{n} a_{ij} = k. \tag{10.1}$$

Let the rows and columns be called 1, 2, ..., n. Construct a bipartite graph G with vertices 1 to n in V_1 and in V_2 and with vertex i in V_1 adjacent to vertex j in V_2 if the ith row of A has unity in the jth column. For example, if

$$A = \begin{bmatrix} 0&0&0&1&1&0&1 \\ 0&1&1&0&1&0&0 \\ 0&0&1&1&0&1&0 \\ 0&1&0&0&0&1&1 \\ 1&1&0&1&0&0&0 \\ 1&0&0&0&1&1&0 \\ 1&0&1&0&0&0&1 \end{bmatrix}, \tag{10.2}$$

vertex 2 in V_1 is adjacent to vertices 2, 3, and 5 in V_2 since for the second row unity occurs in columns 2, 3, and 5.

It follows from (10.1) that G is regular of degree k. Hence, from Theorem 10.6 we can get a 1-factorization of G. Each 1-factor of that factorization corresponds to a permutation matrix, and those permutation matrices sum to A.

Example 10. The matrix A (10.2) corresponds to the regular bigraph of Figure 10.5. The permutation matrices corresponding to the graphs G_1, G_2, G_3 of Figure 10.6 are

$$P_1 = \begin{bmatrix} 0&0&0&1&0&0&0 \\ 0&1&0&0&0&0&0 \\ 0&0&1&0&0&0&0 \\ 0&0&0&0&0&1&0 \\ 1&0&0&0&0&0&0 \\ 0&0&0&0&1&0&0 \\ 0&0&0&0&0&0&1 \end{bmatrix}, \quad P_2 = \begin{bmatrix} 0&0&0&0&1&0&0 \\ 0&0&1&0&0&0&0 \\ 0&0&0&1&0&0&0 \\ 0&0&0&0&0&0&1 \\ 0&1&0&0&0&0&0 \\ 0&0&0&0&0&1&0 \\ 1&0&0&0&0&0&0 \end{bmatrix}, \quad P_3 = \begin{bmatrix} 0&0&0&0&0&0&1 \\ 0&0&0&0&1&0&0 \\ 0&0&0&0&0&1&0 \\ 0&1&0&0&0&0&0 \\ 0&0&0&1&0&0&0 \\ 1&0&0&0&0&0&0 \\ 0&0&1&0&0&0&0 \end{bmatrix}$$

Note that $P_1 + P_2 + P_3 = A$ as desired.

EXERCISES

1. For the Latin rectangle below, find the sets S_1, S_2, S_3, S_4, S_5 used in the proof of Theorem 10.5 and use them to find another row for the rectangle.

$$\begin{matrix} 1&2&3&4&0 \\ 3&1&4&0&2 \\ 4&0&1&2&3 \end{matrix}$$

2. Prove or disprove: given any $k \times n$ Latin rectangle ($k \leqslant n$), and further entries continuing the first l columns ($l < n$) in a Latin way, that collec-

tion of rows and columns can be extended to a Latin square. (See exercise 4, Section 10.1.)

3. For which values of n do the following types of graphs have 1-factors? Cycles C_n, paths P_n (n vertices), complete bigraphs $K_{1,n}$. Which complete bigraphs $K_{m,n}$ have 1-factors?

4. Prove the claim in the text that every complete graph K_{2n} is 1-factorable. [*Hint:* The set of lines $X_i = \{v_i v_{2n}\} \cup \{v_{i-j} v_{i+j}; j = 1, 2, \ldots, n-1\}$ is a 1-factor, if all subscripts are expressed as one of $1, 2, \ldots, (2n-1)$ modulo $(2n-1)$].

5. Write a 5-week schedule for six little league clubs, A, B, C, D, E, and F, to play each other once.

6. Display 1-factorizations of the following graphs.

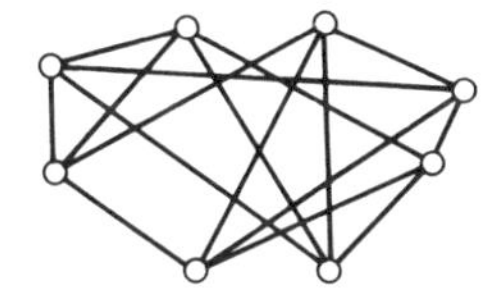
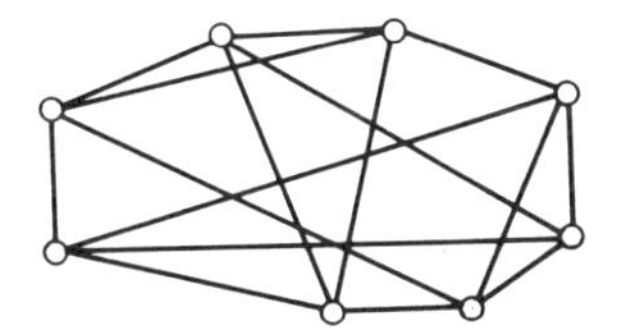

7. Prove that every 1-factorable graph is regular and has an even number of vertices.

8. Write the following matrices as sums of permutation matrices.

$$\begin{bmatrix} 1 & 0 & 0 & 1 & 1 \\ 0 & 1 & 1 & 1 & 0 \\ 0 & 1 & 1 & 0 & 1 \\ 1 & 0 & 0 & 1 & 1 \\ 1 & 1 & 1 & 0 & 0 \end{bmatrix} \qquad \begin{bmatrix} 1 & 1 & 0 & 1 & 0 \\ 0 & 0 & 1 & 1 & 1 \\ 1 & 0 & 1 & 0 & 1 \\ 0 & 1 & 1 & 0 & 1 \\ 1 & 1 & 0 & 1 & 0 \end{bmatrix}$$

9. Prove that if A is a matrix with constant row and column sums and P is a permutation matrix, then $A - P$ has constant row and column sums.

10. Use multigraphs (graphs with multiple edges) to prove that Theorem 10.7 is true for any matrix A with positive integral entries, such that each row and each column sums to k.

11. Write the following matrices as sums of permutation matrices.

$$\begin{bmatrix} 1 & 1 & 1 & 0 & 1 \\ 1 & 0 & 0 & 1 & 2 \\ 1 & 3 & 0 & 0 & 0 \\ 1 & 0 & 3 & 0 & 0 \\ 0 & 0 & 0 & 3 & 1 \end{bmatrix} \qquad \begin{bmatrix} 2 & 2 & 0 & 0 & 0 & 0 & 1 \\ 0 & 1 & 1 & 1 & 0 & 1 & 1 \\ 2 & 0 & 2 & 1 & 0 & 0 & 0 \\ 0 & 1 & 1 & 1 & 1 & 1 & 0 \\ 1 & 0 & 1 & 1 & 1 & 1 & 0 \\ 0 & 1 & 0 & 0 & 2 & 1 & 1 \\ 0 & 0 & 0 & 1 & 1 & 1 & 2 \end{bmatrix}$$

10.4. ALGORITHMS FOR MATCHINGS

What we have done so far with marriages, matchings, and SDR's is mathematically elegant, but it has two serious problems. First, if the conditions of the theorems are not met, then a complete matching, marriage, or SDR is not possible, but it would be nice to know just how close to a complete solution we could come. A matching that is as large as possible is called a *maximum matching*. The second problem is then: How is a maximum matching (complete or not) to be found? Thus far our examples have been so small that it is not too hard to find a matching by examination, but real applications might be far larger and impossible to handle without a systematic approach. We present here an algorithm for finding maximum matchings, complete or not, in bipartite graphs. Obviously the algorithm applies also to marriage problems and SDR's, after an appropriate translation.

What we will use to find maximum matchings is a method that, beginning with any nonmaximum matching, increases the size of that matching. Suppose then we have a graph G with some (probably nonmaximum) matching. Figure 10.7 shows such a G, in which the edges of the matching are bold. A vertex is *free* if it is not on any edge of the matching. Thus the free vertices in Figure 10.7 are v_2, v_5, u_3, u_5, u_6. An *alternating path* is a path whose edges are alternately in the matching and not in it. If an alternating path begins and ends with free vertices it is called an *augmenting path*. The reason that augmenting paths are important is that switching matched and unmatched edges in such

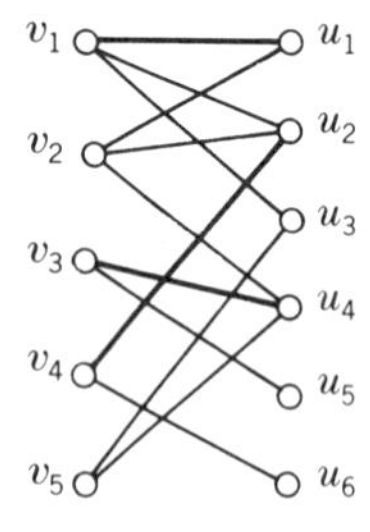

Figure 10.7

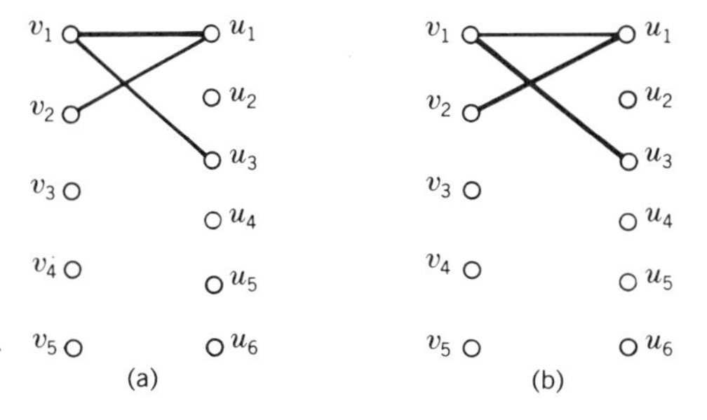

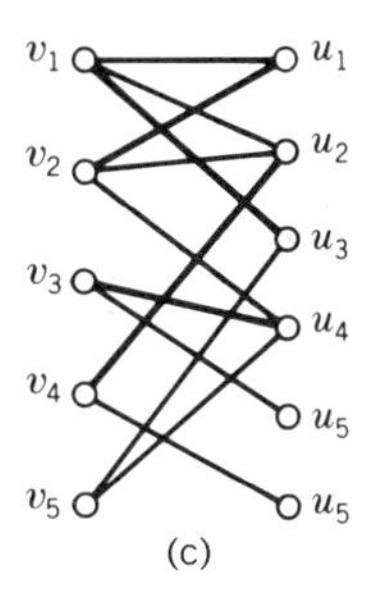

Figure 10.8

a path increases the number of matched edges by 1. In Figure 10.7, $v_2u_1v_1u_3$ is an augmenting path. Switching on that path deletes edge v_1u_1 from the matching and adds the edges v_2u_1 and v_1u_3 as shown in Figure 10.8(a) and (b). The resulting larger matching is shown in Figure 10.5(c).

Thus we have found a way to make matchings larger. Since it is easy to find a matching to start with (the trivial matching with no edges will do), two problems remain. First, does this method always lead to a maximum matching, or might we get stuck short of a maximum? Second, and more basic, how can augmenting paths be found? We consider the questions in turn.

The method of augmenting paths always does lead to a maximum matching because of the following theorem.

Theorem 10.8. If a matching E is not maximum, then there is an augmenting path for E.

Proof. Suppose E is the set of edges that make up the matching and F is the set of edges of some maximum matching. We need to show that if $|F| > |E|$, the matching E has an augmenting path. Define the sum $E \oplus F$ to be those edges that are in E or F but not in both. Thus $E \oplus F = (E \cup F) - (E \cap F)$. Since E and F are both subgraphs that are regular of degree 1, each vertex is touched by at most 2 edges of $E \oplus F$. In fact, if a vertex is touched by 2 edges of $E \oplus F$, one must be from E and the other from F. Thus $E \oplus F$ consists entirely of paths and cycles in which the edges alternate between E and F. Each piece of $E \oplus F$ looks like one of the cases shown in Figure 10.9. Typical paths, with more edges from F, more edges from E, and equal numbers of edges from E and F are shown in (a), (b), and (c). A typical cycle, which has equal numbers of E and F edges, is shown in (d). Now remember that we are supposing that $|F| > |E|$ since F is maximum but E is not. When $E \oplus F$ is formed, there is some cancellation of edges, but equal numbers of

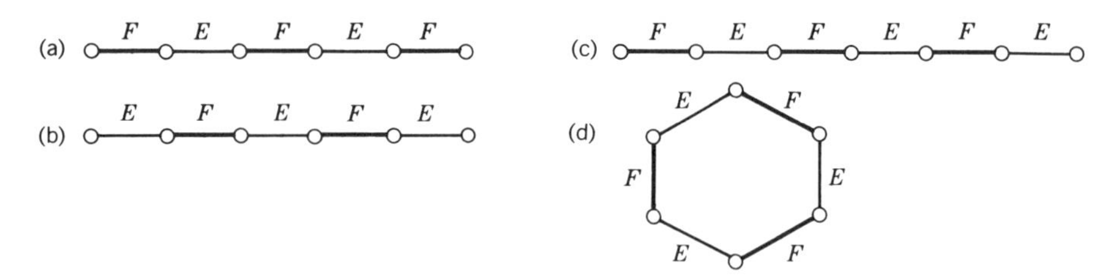

Figure 10.9

edges are lost from E and F. Thus $E \oplus F$ contains more edges from F than from E. But (c) and (d) of Figure 10.9 are pieces of $E \oplus F$ that have equal numbers of E and F edges, while paths such as these shown in Figure 10.9(b) actually have more E edges than F edges. So if $E \oplus F$ has more edges from F than from E, $E \oplus F$ must contain at least one path such as the one shown in Figure 10.9(a), which means E has an augmenting path. This proves the theorem.

Now we know that augmenting paths are a way to find maximum matchings. What we do not have is a means for finding augmenting paths. For that, we present the following algorithm.

Algorithm for Augmenting Paths

Given: A bipartite graph G with vertex sets V_1 and V_2 and a matching (perhaps the empty matching).

Then: Find an augmenting path for the matching or decide that no augmenting path exists (so the matching is maximum).

Method: (1) Label any unlabeled free vertex in V_1 with the label (*) indicating the start of an alternating path.

(2) If $v \in V_1$ is labeled, $u \in V_2$ is unlabeled, and uv is an unmatched edge of G, then label u with label (v).

(3) If $v \in V_1$ is unlabeled, $u \in V_2$ is labeled, and uv is a matched edge of G, label v with label (u).

(4) Repeat (1), (2), and (3) above until either

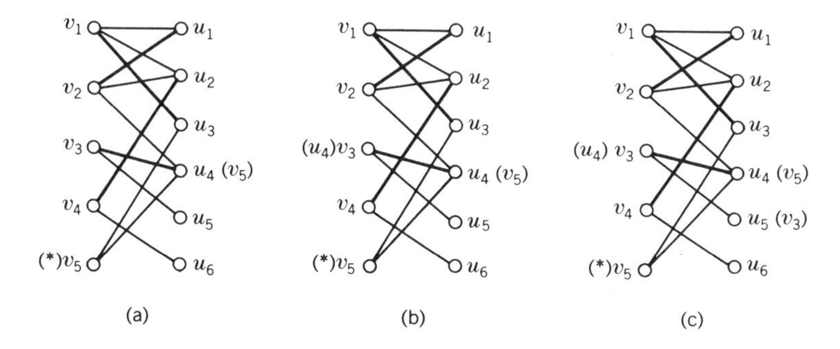

Figure 10.10

(i) a free vertex $u \in V_2$ is labeled or
(ii) nothing more can be labeled.

We claim that in case (i) an alternating path has been found and in case (ii) G contains no alternating path. An example will make the first claim clear.

Example 11. In Figure 10.10(a) we follow rule (1) to label vertex v_5 in V_1 with (*) and, following the unmatched edge v_5u_4, we label vertex u_4 with (v_5) as prescribed by rule (2). In part (b), we follow the matched edge u_4v_3 and label vertex v_3 with (u_4). Finally, in (c) we follow the unmatched v_3u_5 and label vertex u_5 with (v_3). Since u_5 is a free vertex, we have found an alternating path. That path can be recovered by starting with vertex u_5 and tracing back by labels to v_3, u_4, and v_5. When vertex v_5 is found with a (*), we know exactly where the path $v_5u_4v_3u_5$ begins.

The algorithm we have outlined can be run either in depth-first fashion, labeling a single path until stuck and back-tracking to other possibilities, or in a breadth-first fashion, labeling all free vertices in V_1 with (*), then labeling all vertices adjacent to those along unmatched edges, and so on. We ran the example in depth-first fashion because that minimized the number of labels and therefore the confusion. An actual implementation would probably run in breadth-first fashion instead.

In any case, the algorithm should be made to run through all possible sequences of labelings. We need to show that if there is an augmenting path, this method of labelling will find it. Suppose there is an augmenting path P that begins at x. Then clearly x is free, so the algorithm labels $x(*)$. There is clearly a last vertex on P that is labeled by the algorithm. Call that vertex w. Then w must be the free vertex at the end of P because otherwise either rule (1) or rule (2) could be applied to label the next vertex on P.

EXERCISES

1. Which vertices are free in Figure 10.8(c)?
2. Find all alternating paths beginning at vertex v_1 in Figure 10.7. Which of those paths are augmenting?
3. Find four augmenting paths for the graph of Figure 10.8(c) and display the four matchings that result from switching on each of them.
4. Call the matching of Figure 10.8(c) E and the following maximum matching (on the same graph) F. Draw the sum $E \oplus F$ for this E and F, and verify that it is a path with edges alternating between E and F, beginning and ending with edges of F.

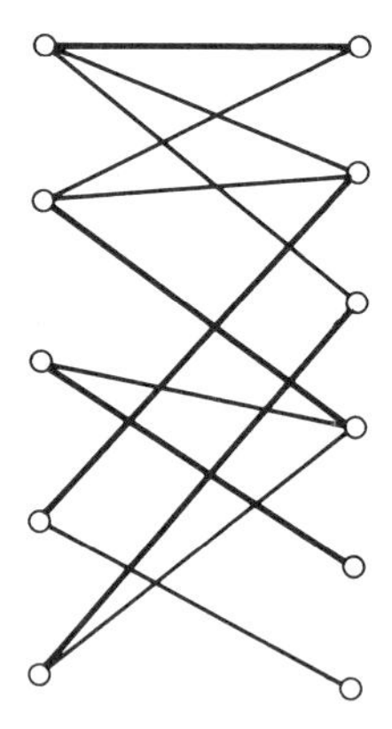

5. Run the algorithm for alternating paths on the graph of Figure 10.8(c) as follows:
   (a) In depth-first fashion, always choosing the edge toward the top of the graph, if there is a choice.
   (b) In breadth-first fashion.
   In each case, show all labels obtained along the way.

10.5. MAXIMUM FLOWS

A different kind of augmenting path is commonly used for finding maximum flows in networks.

A *network* is a directed graph in which there are two distinguished vertices,

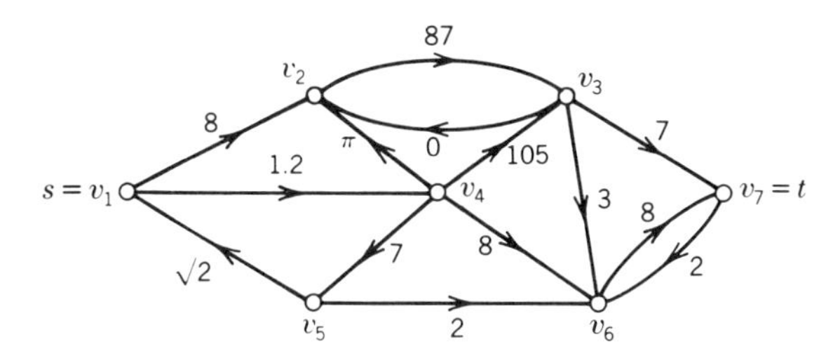

Figure 10.11

a *source* s and a *sink* t, and each edge is assigned a non-negative number called its *capacity*. For convenience, the vertices of the network are labeled with integers and we denote the capacity of the edge from vertex v_i to vertex v_j by $c(i, j)$. Figure 10.11 shows a network with vertex v_1 as source, and vertex v_7 as sink.

Networks are used to investigate flows from one point to another. Examples include transport of goods over a railway or highway system, flow of oil through interconnecting pipes, and flow of electricity through a transmission system or an electronic device.

A *flow* in a network is a function assigning a non-negative number $f(i, j)$ to each edge v_iv_j so that $f(i, j) \leqslant c(i, j)$ and $\Sigma_i f(i, j) = \Sigma_k f(j, k)$, for each vertex v_j that is not s or t. This formal definition of a flow coincides with our intuitive idea of what a flow should be. The first condition assures that the amount of material moving along a particular edge is within the capacity of the edge, while the second assures that at every vertex between the source and sink exactly the same amount of material comes into and goes out of the vertex. No restriction is put on the total net flow out of s or into t, but it is not hard to show that the net flow out of s must equal the net flow into t. This common value is called the *value* of the flow. A *maximum flow* in a network is a flow that has the largest possible value. The principal problem of network flow theory is to determine the largest possible flow value for a given network and to find a flow achieving that value.

A *cut* in a network N is a partition of the vertices of N into two disjoint sets S and T with $s \in S$ and $t \in T$. The capacity of the cut (S, T) is the sum of the capacities of all edges from S to T. Thus, if $c(S, T)$ is the capacity of (S, T),

$$c(S, T) = \sum_{v_i \in S} \sum_{v_j \in T} c(i, j).$$

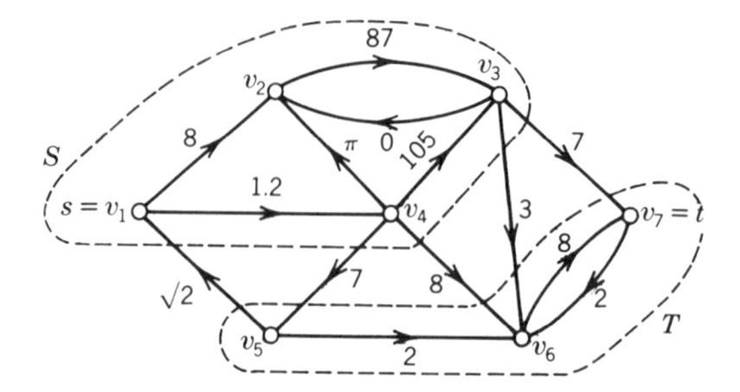

Figure 10.12

Figure 10.12 repeats the network of Figure 10.11 and displays a cut with $S = \{v_1, v_2, v_3, v_4\}$ and $T = \{v_5, v_6, v_7\}$. The capacity of the cut (S, T) is $7 + 8 + 3 + 7 = 25$. Note that the edge v_5v_1 with capacity $\sqrt{2}$ is not included in this sum, since it goes from T to S rather than from S to T.

The max-flow min-cut theorem of network flow theory, first stated by Ford and Fulkerson, is that the largest flow has exactly the same magnitude as the smallest cut. In order to prove that, we need to first understand how a given flow can be increased. Figure 10.13 shows a simple network with capacity and flow indicated on each edge. Edges such as v_2v_3, v_2v_4 are said to be *saturated* because the flow uses their entire capacity. Clearly the indicated flow, with value 9, is not maximum, because the edges v_1v_4 and v_4v_5 are not saturated. Thus the path $v_1v_4v_5$ can have flow increased by 1 on each edge, producing the situation shown in Figure 10.13(b). The flow now has value 10, but it is still not maximum. It turns out that undirected path $v_1v_4v_3v_5$ can be used to increase the flow by increasing the flow on v_1v_4 and v_3v_5 by 1 and decreasing the flow on v_3v_4 by 1. These changes yield the network of Figure 10.13(c) with flow 11, which is maximum.

The changes we have made to move from the original flow to a maximum one in Figure 10.13 suggest a method that will always yield a maximum flow. If P is an undirected path from s to t we call an edge of P *forward* if its

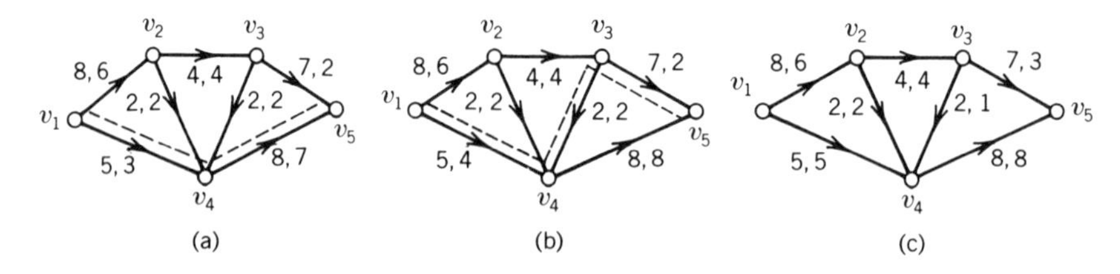

Figure 10.13

direction on P is from s to t and *backward* otherwise. Given a flow f, P is a *flow augmenting path* if $f(i, j) < c(i, j)$ for each forward arc v_iv_j of P and $f(i,j) > 0$ for each backward arc v_iv_j. Thus forward arcs are unsaturated and backward arcs have some flow along them. If f has a flow augmenting path, then we can increase the value of the flow by increasing flow on forward arcs and decreasing it on backward arcs, by an amount d. Here d is the minimum of the values $c(i, j) - f(i, j)$ [over forward arcs (i, j)] and $f(k, m)$ [over backward arcs (k, m)]. Since all of these quantities are positive for a flow-augmenting path, and there are only finitely many of them, d is positive. Note that such changes preserve the balance of flows into and out of each vertex besides s and t, and increase the value of the flow by d.

Thus we have a method that will find large flows; begin with any flow, find an augmenting path, use that path to increase the flow, and repeat the process. We now present an effective way to find augmenting paths that both finds maximum flows and gives a way to prove the max-flow min-cut theorem.

Algorithm for Flow Augmenting Paths

Given: A network N, with source s and sink t, and a (possibly zero) flow from s to t.

Find: A flow augmenting path from s to t, or decide that no such path exists (so the flow is maximum).

Method:
(1) Label $s(\phi, \infty)$ indicating no predecessor and infinite capacity.
(2) If vertex v is labeled (a, x), where a is arbitrary *and* there is an unsaturated edge from v to w with capacity c and flow f, *and* w is unlabeled, then label w with the pair $(v, \min(x, c - f))$.
(3) If vertex v is labeled (a, x), where a is arbitrary *and* there is an edge with positive flow f from w to v, *and* w is unlabeled, then label w with the pair $(v, \min(x, f))$.
(4) Repeat (2) and (3) above until either
  (i) t is labeled or
  (ii) no more vertices can be labeled.

It is clear that in case (i) we have found a flow-augmenting path from s to t. That path can be retraced using the first half of each vertex label. The amount by which the flow can be increased is given by the second half of t's label Repeating the process until we are stuck, we claim that in case (ii) there is no flow-augmenting path from s to t and moreover the flow we have is then maximum and equal in value to the size of a minimum cut.

Obviously, a maximum flow cannot be greater than a minimum cut. Suppose that t is not labeled by the algorithm, so we have case (ii). If X is the set of vertices that are labeled and $Y = V - X$, then s is in X and t is in Y. Since the algorithm cannot label any more vertices, by rule (2) all edges from X to Y must have flow equal to their capacity. Moreover, all edges from Y to X must have zero flow, or rule (3) of the algorithm would apply. But this implies that the flow from X to Y, which is clearly equal to the flow from s to t, is exactly equal in size to the capacity of the cut (X, Y). Since we have a flow equal to a cut, the maximum flow is at least as large as the minimum cut. Thus we have proved the following theorem.

Theorem 10.9. (Max-Flow Min-Cut) For any network the maximum flow is equal in value to the minimum cut.

The algorithm we have given has a serious disadvantage: If the capacities are not rational, it may never terminate. Thus for certain networks it may keep finding augmenting paths, but the increase in flow with each path is so small that the maximum flow is never found. Even for nice capacities, the algorithm may be very inefficient. Figure 10.14 shows a nice small network that obviously has maximum flow 2000. If, however, the algorithm happens to begin with zero flows and finds the augmenting paths $v_1v_3v_2v_4$ and $v_1v_2v_3v_4$, alternately, then it will take 2000 iterations to arrive at that maximum flow. There are fairly simple modifications of the algorithm that avoid such problems and find the maximum flow in a reasonable length of time, not only for the network of Figure 10.14 but also for networks with irrational capacities.

There are variations of the max-flow min-cut theorem that deserve mention. Instead of having a limited capacity on each edge, it is not hard to imagine applications in which the capacity of each vertex (an intersection, a switching station) is limited. In such a case, a *vertex cut* is clearly a set of vertices whose removal destroys all directed paths from s to t, and the value of the cut is the sum of the capacities of those vertices.

Theorem 10.10. For any network N with vertex capacities, the maximum flow is equal in value to the minimum vertex cut.

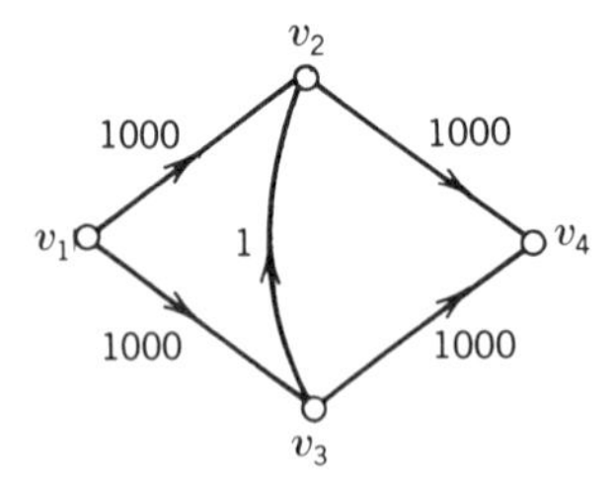

Figure 10.14

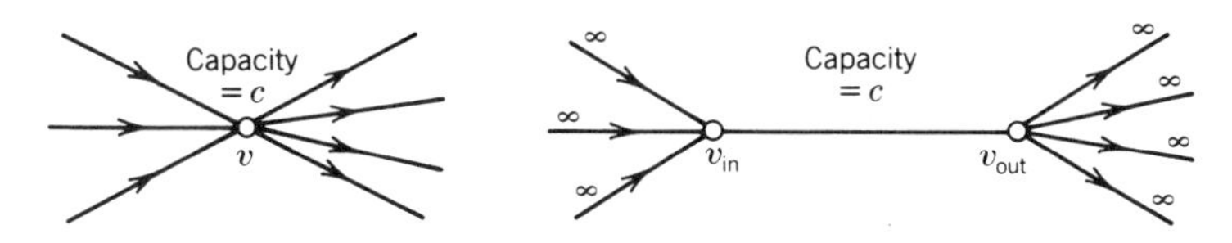

Figure 10.15

Proof. Given N, we construct a new network N^* by converting each vertex v with capacity c into two vertices v_{in} and v_{out}, joined by an arc of capacity c. The arcs into v come to v_{in} and the arcs out of v leave v_{out} (see Figure 10.15). The arcs of the original network are all given capacity ∞. Then N^* is a network with edge capacities and the max-flow min-cut theorem applies. Obviously, no edges with capacity ∞ are in a minimum cut, so the minimum cut corresponds to a vertex cut in N and the value of that vertex cut is equal to the value of the maximum flow in N^*, which corresponds to a maximum flow in N.

The method of Theorem 10.10 can clearly also be applied to determine flows in networks in which capacities are specified both for vertices and for arcs. The special case of Theorem 10.10 in which all of the vertex capacities are one was proved by Menger in 1927, long before the max flow min cut theorem was conceived.

Corollary 10.2. (Menger's Theorem) In any graph G the maximum number of disjoint paths joining any two nonadjacent vertices s and t is equal to the minimum number of points whose removal separates s and t in G.

Although Menger's Theorem is a special case of a variation of max-flow min-cut, it is powerful. We can easily prove Hall's Theorem from it, for example. Suppose that the subsets $S_1, S_2, \ldots, S_n$ of a set S have the required property for an SDR: The union of any k sets contains at least k distinct elements of S. Then we form a bipartite graph G with $V_1 = \{S_1, S_2, \ldots, S_n\}$, $V_2 = S = \{x_1, x_2, \ldots, x_m\}$, and $S_i x_j$ an edge if $x_j \in S_i$. From G we form a new graph G^* that has a new vertex s adjacent to all vertices of V_1 and a new vertex t adjacent to all vertices of V_2, as in Figure 10.16. Clearly, the graph G^* has n disjoint paths from s to t if and only if the sets $S_1, S_2, \ldots,$ S_n have an SDR. But we know from Menger's Theorem that there are n disjoint paths if the minimum set of vertices separating s from t is of size n. Suppose that such a minimum separating set M contains n_1 vertices from V_1 and n_2 vertices from V_2. Then there are $n - n_1$ vertices in V_1 that are not in M and, by the condition of Hall's Theorem, those vertices are joined to at least $n - n_1$ vertices of V_2. Because M separates s from t, all of those $n - n_1$ vertices

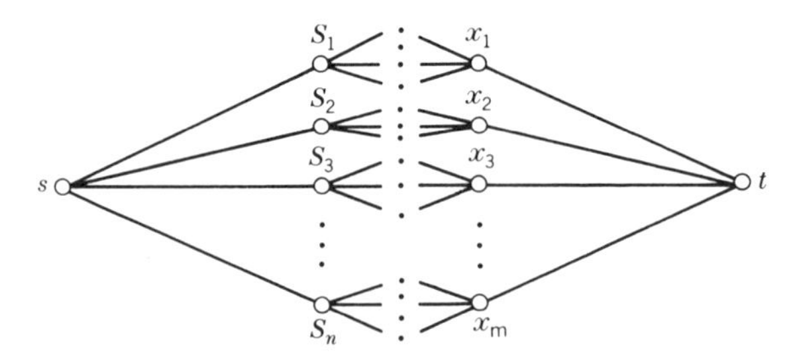

must be in $V_2 \cap M$, which has size n_2, so $n_2 \geqslant n - n_1$. Thus $n_1 + n_2 \geqslant n$, and we have shown that a minimum separating set of vertices is of size at least n. Since V_1 separates, a minimum separating set is of size exactly n. Thus there are n disjoint paths from s to t that yield an SDR, and Hall's Theorem is true.

A similar method can be used to prove a theorem on 0–1 matrices. A *line* of matrix is a row or a column. Two entries are *independent* if they do not lie in a line, and a set of entries is independent if each pair of its elements is an independent pair.

Theorem 10.11. For any 0–1 matrix the maximum size of an independent set of unit entries equals the minimum number of lines required to cover all unit entries.

Proof. Say $A = [a_{ij}]$ is an $n \times m$ 0–1 matrix and form a bipartite graph G that has $V_1 = \{x_1, x_2, \ldots, x_n\}$, $V_2 = \{y_1, y_2, \ldots, y_m\}$. In G, $x_i y_j$ is an edge if and only if $a_{ij} = 1$. Then augment G by adjoining s adjacent to the vertices of V_1 and t adjacent to the vertices of V_2 and call the resulting graph G^*. Each line in G^* between V_1 and V_2 corresponds to a unit entry in A, and two unit entries are independent if and only if the corresponding lines do not share an end vertex. Thus a maximum set of disjoint paths from s to t, which gives a maximum matching from V_1 to V_2, corresponds to a maximum independent set of units in A. Say the size of such a set is max. On the other hand, the unit entries in a row of A correspond to the lines to V_2 from one vertex in V_1 and the entries in a column of A correspond to lines to V_1 from some vertex in V_2. Thus the minimum number of lines required to cover all units in A equals the minimum number of vertices required to separate s and t by destroying all edges from V_1 to V_2. Call that number min. Menger's Theorem tells us that max = min, so the theorem is proved.

Example 12. Seven men are qualified for ten jobs as shown in the following matrix:

Man \ Job	A	B	C	D	E	F	G	H	I	J
1	0	1	0	1	0	0	0	1	0	0
2	0	0	0	1	0	1	0	0	0	0
3	0	1	0	1	0	0	0	0	0	0
4	0	0	0	0	0	1	0	0	1	1
5	0	1	0	0	0	0	0	1	0	0
6	0	0	0	0	0	1	0	1	0	0
7	1	0	1	0	1	0	1	0	0	0

Find an assignment of jobs that maximizes the number of men working.

Since each man can hold only one job, and each job only employs one man, an assignment is a set of independent units. It is not difficult to find a way to assign six men to different jobs. To see that seven men cannot be so assigned, note that rows 4 and 7 and columns B, D, F, and H together cover all units, so the maximum independent set is of size six.

Example 12 is a restatement of Example 1, on boys and girls, with which we began the chapter. At this point we not only have a theorem that tells us just how many men can be employed, but by recasting the problem in terms of matchings in bipartite graphs, we have an alternating path algorithm for finding a maximum assignment. What we have covered in this chapter is indeed just one part of the vast area of combinatorial optimization.

EXERCISES

1. On the three networks below, each edge has been labeled with a capacity and a "flow." For each network either state why the numbers shown do not constitute a flow or state the value of the flow.

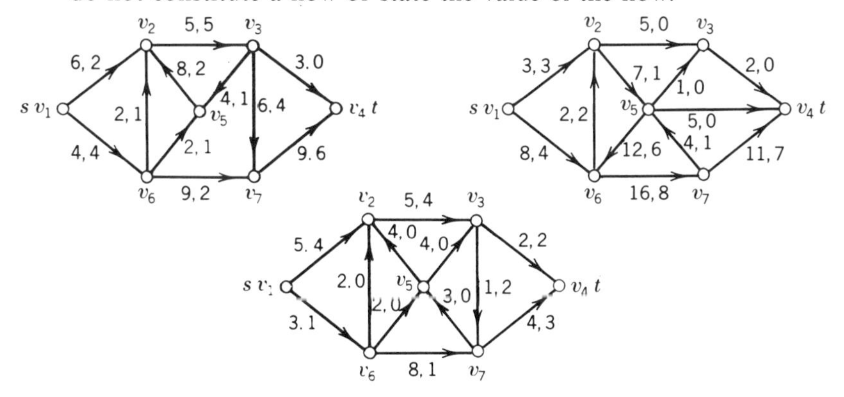

2. Prove the claim in the text that for every flow f, the net flow out of s must equal the net flow into t.

3. Prove that for any network with positive integral capacities there exists an integral flow function that is maximum.

4. Find all possible integral flows f for the network shown below with indicated capacities.

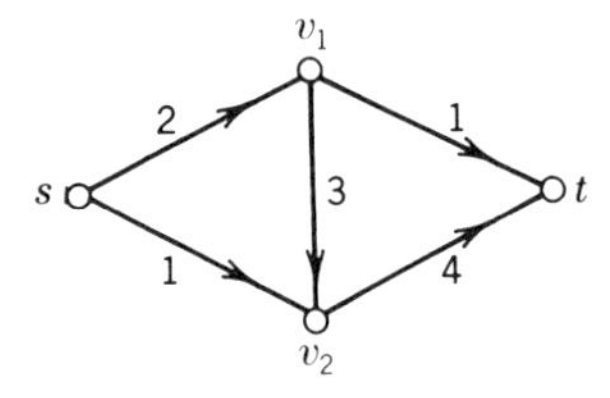

5. Find a flow-augmenting path in each of the two networks below, determine the largest possible augmenting number d for each path, and exhibit the augmented flows.

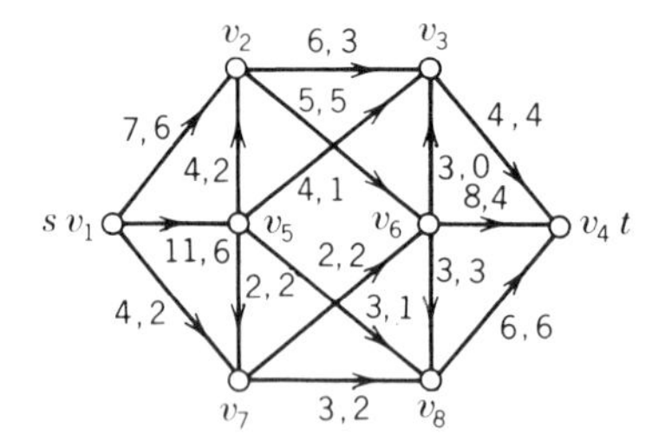

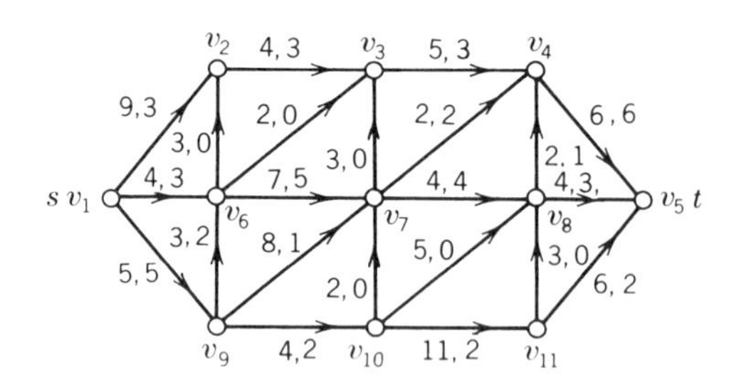

6. Verify the max-flow min-cut theorem for the networks of exercise 5 by exhibiting for each one a flow and a cut of equal value.

7. Is it necessarily true that for a maximum flow f, $f(i, j) \cdot f(j, i) = 0$ for all i and j?

8. Prove or disprove each half of the double implication: An edge is an edge of greatest capacity in some minimum cut set, if and only if its removal from the network decreases the maximum s–t flow by at least as much as the deletion of any other edge.

9. Various sequences of flow-augmenting paths can be applied to the network shown below to increase the flow from the zero flow to the maximum flow. What is the smallest number of augmenting paths which could be used? The largest? Explain.

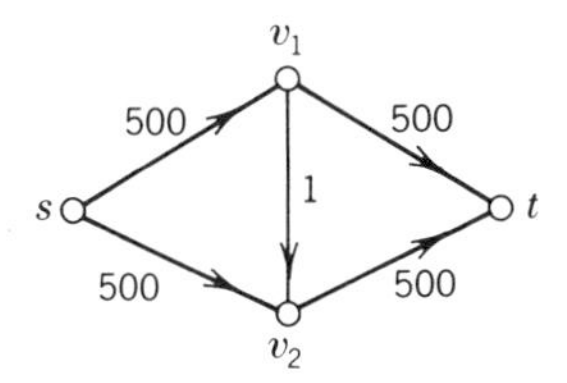

10. Apply the max-flow algorithm to the network below, beginning with the zero flow. Show the labels at each stage.

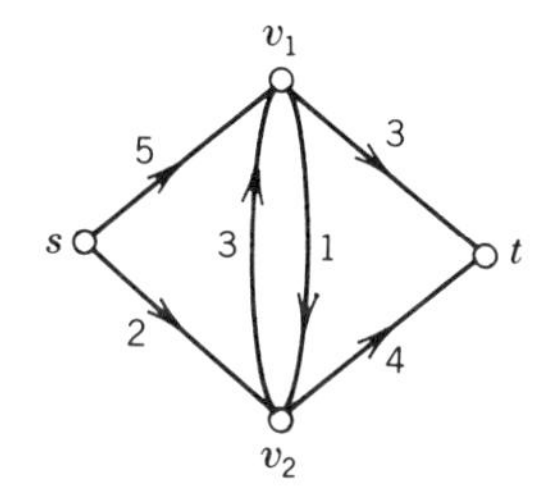

11. Suppose a network N has n vertices, each with a vertex capacity. How many vertices does the corresponding network N^*, with edge capacities, have?

12. For the network N below with vertex capacities:

(a) Draw the corresponding network N^* with capacities on edges.

(b) Find the maximum S–T flow.

(c) Find a minimum vertex cut in N.

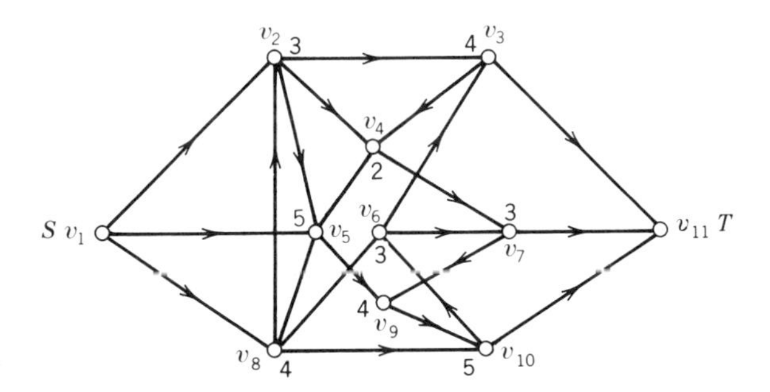

13. State and prove a general theorem relating the maximum flow from a set of vertices U to a set of vertices V in a network N. An example of such a network would be:

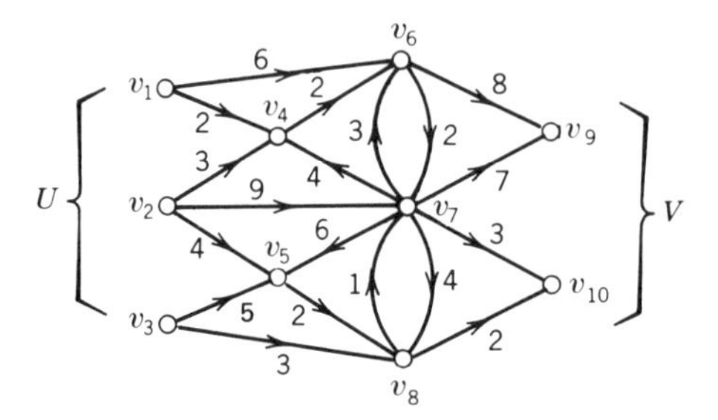

14. Verify Menger's Theorem for the following graphs.

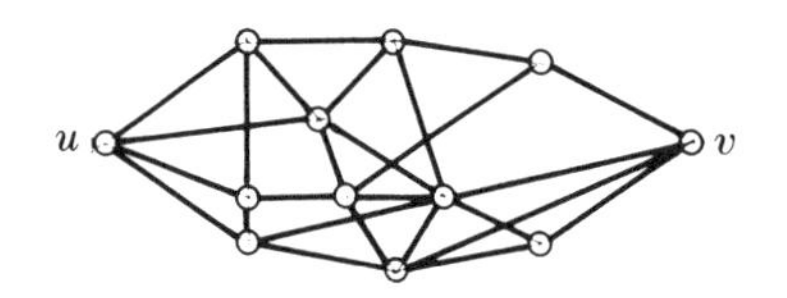

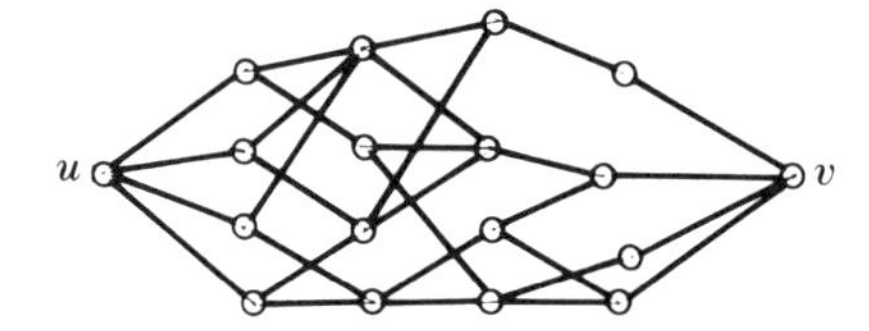

15. (a) Draw a graph (network) corresponding to Example 1 of this chapter, with boys all connected to s, girls all connected to t, and boys and girls connected as they are acquainted.

(b) In the graph of part (a), find a set of six vertices that separate s from t. What does that show?

16. For each of the following 0–1 matrices find a maximum independent set of units, and prove that it is maximum.

$$\begin{bmatrix} 0&1&0&1&1&0&1\\ 1&0&0&1&0&0&0\\ 1&0&0&1&1&0&0\\ 0&1&0&1&1&0&0\\ 0&1&0&0&1&0&1\\ 0&0&1&0&1&1&1\\ 1&1&0&1&0&0&1 \end{bmatrix}$$

(a)

$$\begin{bmatrix} 0&1&0&0&1&0\\ 0&0&0&1&0&0\\ 1&0&1&1&0&1\\ 0&1&0&0&1&0\\ 0&1&0&1&1&0\\ 0&1&1&1&0&1 \end{bmatrix}$$

(b)

$$\begin{bmatrix} 0&1&0&0&0&0\\ 0&0&0&0&1&0\\ 1&1&1&0&0&1\\ 0&0&0&0&1&0\\ 0&1&1&1&1&1\\ 0&1&0&0&0&0 \end{bmatrix}$$

(c)

10.6. Remarks

The material covered in this chapter first arose in diverse areas of mathematics. Menger proved his theorem on disjoint paths in graphs (Corollary 10.2) in 1927. Theorem 10.11, on 0–1 matrices, was proved by König and Egervary in 1931, and König derived many fundamental theorems on factorization of graphs in his pioneering graph theory book (1936). Meanwhile, Phillip Hall proved Theorem 10.3 on systems of distinct representatives (also frequently called *transversals*) in 1935. Thus, the theoretical base of this branch of combinatorial optimization appeared in a rush, about fifty years ago.

Twenty years later, in 1956, the max-flow min-cut theorem (our Theorem 10.9) was published simultaneously by Ford and Fulkerson and Elias, Feinstein, and Shannon. (This theorem is most often associated with Ford and Fulkerson because of their very important book on networks [1].) In 1950, Halmos and Vaughan presented the proof of the marriage theorem that we use here. Interest in effective methods for matchings and flows increased, as the tools became available for implementing such methods, and combinatorial optimization has grown to be a branch of mathematics in its own right.

Probably the most complete book on combinatorial optimization accessible to those not versed in operations research is by Lawler [2]. We have excluded all discussion of efficiency, which he covers extensively. Lawler also presents the mathematical roots of optimization in the theory of matroids. A very short introduction to matroids can be found in Wilson [3].

[1] L. R. Ford and D. R. Fulkerson, *Flows in Networks*, Princeton University Press, Princeton, 1962.

[2] E. L. Lawler, *Combinatorial Optimization: Networks and Matroids*, Holt, Rinehart and Winston, New York, 1976.

[3] R. J. Wilson, *An Introduction to Graph Theory*, Academic Press, New York, 1972.

Index

Applied Probability and Statistics (Continued)

DRAPER and SMITH • Applied Regression Analysis, *Second Edition*
DUNN • Basic Statistics: A Primer for the Biomedical Sciences, *Second Edition*
DUNN and CLARK • Applied Statistics: Analysis of Variance and Regression
ELANDT-JOHNSON and JOHNSON • Survival Models and Data Analysis
FLEISS • Statistical Methods for Rates and Proportions, *Second Edition*
FOX • Linear Statistical Models and Related Methods
FRANKEN, KÖNIG, ARNDT, and SCHMIDT • Queues and Point Processes
GALAMBOS • The Asymptotic Theory of Extreme Order Statistics
GIBBONS, OLKIN, and SOBEL • Selecting and Ordering Populations: A New Statistical Methodology
GNANADESIKAN • Methods for Statistical Data Analysis of Multivariate Observations
GOLDBERGER • Econometric Theory
GOLDSTEIN and DILLON • Discrete Discriminant Analysis
GREENBERG and WEBSTER • Advanced Econometrics: A Bridge to the Literature
GROSS and CLARK • Survival Distributions: Reliability Applications in the Biomedical Sciences
GROSS and HARRIS • Fundamentals of Queueing Theory
GUPTA and PANCHAPAKESAN • Multiple Decision Procedures: Theory and Methodology of Selecting and Ranking Populations
GUTTMAN, WILKS, and HUNTER • Introductory Engineering Statistics, *Third Edition*
HAHN and SHAPIRO • Statistical Models in Engineering
HALD • Statistical Tables and Formulas
HALD • Statistical Theory with Engineering Applications
HAND • Discrimination and Classification
HILDEBRAND, LAING, and ROSENTHAL • Prediction Analysis of Cross Classifications
HOAGLIN, MOSTELLER, and TUKEY • Understanding Robust and Exploratory Data Analysis
HOEL • Elementary Statistics, *Fourth Edition*
HOEL and JESSEN • Basic Statistics for Business and Economics, *Third Edition*
HOGG and KLUGMAN • Loss Distributions
HOLLANDER and WOLFE • Nonparametric Statistical Methods
IMAN and CONOVER • Modern Business Statistics
JAGERS • Branching Processes with Biological Applications
JESSEN • Statistical Survey Techniques
JOHNSON and KOTZ • Distributions in Statistics
Discrete Distributions
Continuous Univariate Distributions—1
Continuous Univariate Distributions—2
Continuous Multivariate Distributions
JOHNSON and KOTZ • Urn Models and Their Application: An Approach to Modern Discrete Probability Theory
JOHNSON and LEONE • Statistics and Experimental Design in Engineering and the Physical Sciences, Volumes I and II, *Second Edition*
JUDGE, HILL, GRIFFITHS, LÜTKEPOHL and LEE • Introduction to the Theory and Practice of Econometrics
JUDGE, GRIFFITHS, HILL and LEE • The Theory and Practice of Econometrics
KALBFLEISCH and PRENTICE • The Statistical Analysis of Failure Time Data